Graphing Functions Using Transformations

for Algebra and Pre-Calculus

By Kathryn Paulk
Copyright © 2022

Updated: 02/01/2022

TABLE OF CONTENTS

INTRODUCTION 4

GRAPHING TOOLS 6

    PARENT FUNCTIONS 7

    TRANSLATIONS 8

EQUATIONS TO GRAPHS 11

    LINEAR FUNCTIONS 13

    QUADRATIC FUNCTIONS 14

    CUBIC FUNCTIONS 15

    EXPONENTIAL FUNCTIONS 16

    ABSOLUTE VALUE FUNCTIONS 17

    SQUARE ROOT FUNCTIONS 18

    LOGARITHMIC FUNCTIONS 19

    RATIONAL FUNCTIONS (PART 1) 20

    RATIONAL FUNCTIONS (PART 2) 21

    STEP FUNCTIONS (GREATEST INTEGER <= X) 22

    SINE FUNCTIONS 24

    COSINE FUNCTIONS 26

    TANGENT FUNCTIONS 28

GRAPHS TO EQUATIONS.. 29

    LINEAR FUNCTIONS.. 30

    QUADRATIC FUNCTIONS 31

    CUBIC FUNCTIONS 32

    EXPONENTIAL FUNCTIONS 34

    ABSOLUTE VALUE FUNCTIONS 35

    SQUARE ROOT FUNCTIONS 36

    LOGARITHMIC FUNCTIONS 37

    RATIONAL FUNCTIONS (PART 1).................................. 38

    RATIONAL FUNCTIONS (PART 2).................................. 39

    STEP FUNCTIONS (GREATEST INTEGER <= X) 40

    SINE FUNCTIONS 41

    COSINE FUNCTIONS 42

    TANGENT FUNCTIONS 43

INTRODUCTION

This book shows how to (1.) sketch a graph for a given function and (2.) how to find the equation of a function for a given graph. Two sets of examples, from simple to complex, demonstrate the concepts. This book is essentially a picture book that demonstrates graphing functions, with a set of examples.

For all examples, the given function or graph is compared to a similar, basic function (parent function). Then, translations (vertical and horizontal) are identified. Applying the translations to the basic function transforms it into the given function.

Students often learn how to graph linear functions by plotting two points on a graph and drawing a straight line through them. For quadratic functions, students may plot three points and draw a parabola through them. Initially, students learn to find points for a given function, then sketch the graph.

As functions become more complex identifying and using translations becomes necessary. It is helpful to use translations on simple functions (e.g. linear) before using them on the more complex functions (e.g. trigonometric).

The examples in this book include the following function types:

- Linear
- Quadratic
- Cubic
- Exponential
- Absolute Value
- Square Root
- Logarithmic
- Rational
- Step (Greatest Integer or Floor)
- Sine
- Cosine
- Tangent

Two sets of one-page examples are included in this book. The first set includes examples for making a graph for a given function. The second set includes examples for finding an equation for a given graph. Parent functions and translations are used with all examples.

GRAPHING TOOLS

Familiarity with basic, parent functions and translations is helpful when graphing and transforming functions. The following sections are summaries of parent functions and translations, from my book "One-Page Summaries for Algebra, Geometry, and Pre-Calculus."

Parent Functions

This summary is from my book "One-Page Summaries for Algebra, Geometry, and Pre-Calculus."

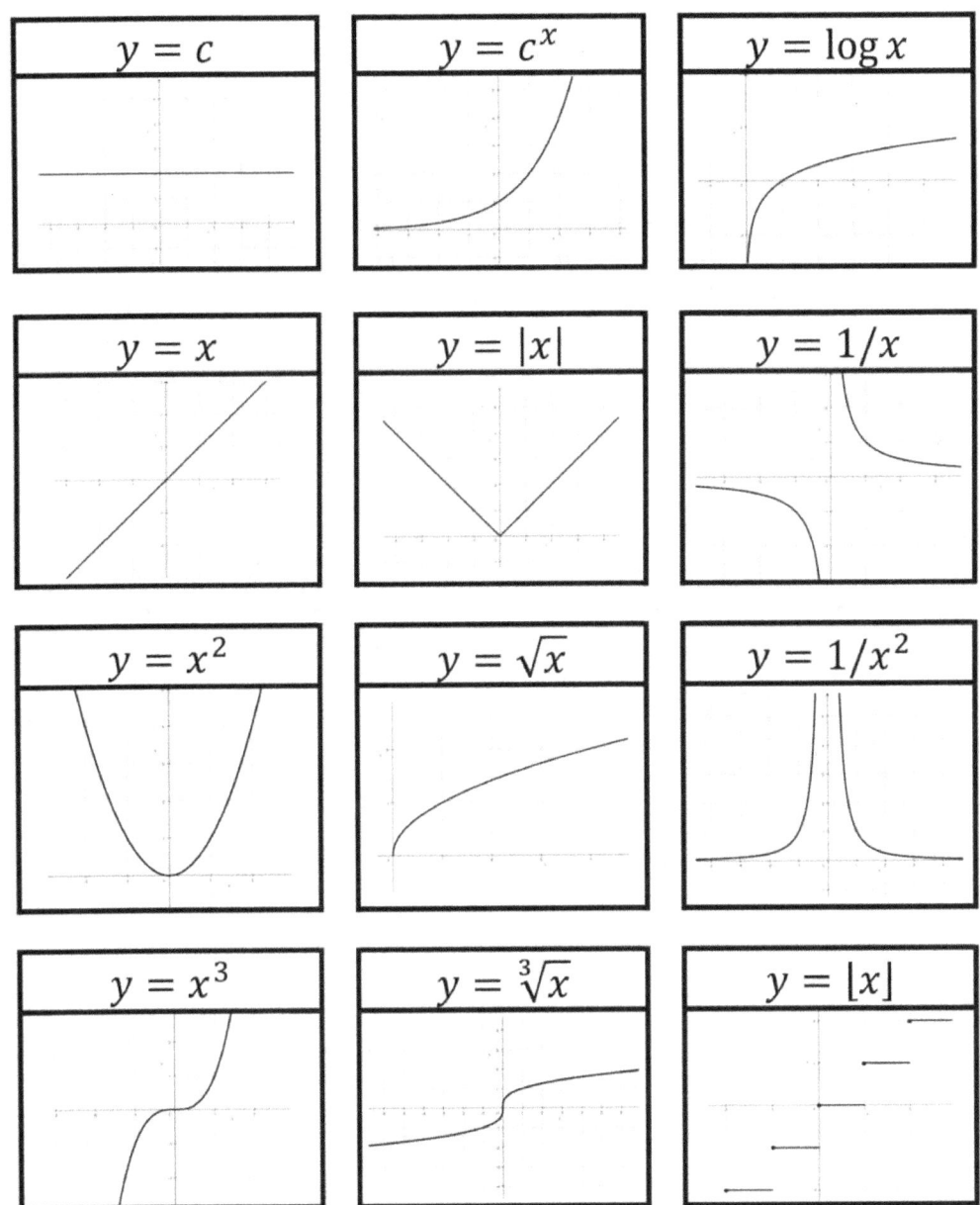

Translations

This summary is from my book "One-Page Summaries for Algebra, Geometry, and Pre-Calculus."

Parent Function	(Example)
$f(x) = x^2$	

Vertical Translations			(Very Nice)
$y = f(x) + a$	Shift Up	↑	
$y = a * f(x)$	Stretch	↕	
$y = -f(x)$	V. Rotation	↷	

Horizontal Translations			(Horrible)
$y = f(x + a)$	Shift Left	←	
$y = f(a * x)$	Compression	→←	
$y = f(-x)$	H. Rotation	↷	

There are two types of translations – vertical translations and horizontal translations.

Vertical translations are "Very Nice" because they are what you would expect. If you add 4 to a function, you would expect the function to increase by 4 and that is exactly what happens. If you multiply a function by 4, you would expect to make it bigger and that is exactly what happens -- it does a vertical stretch by 4. Likewise for subtraction and division – exactly what you would expect.

Horizontal translations are "Horrible" because they are NOT what you would expect. If you add 4 to x then you would expect the function to shift to the right, but that is not what happens. The function shifts horizontally to the left by 4. If you multiply x by 4 then you would expect a horizontal stretch. But, that is not what happens. Multiplying x by 4 actually compresses the function horizontally. It is horrible!

Many students find it difficult to classify a translation as vertical or horizontal. Thinking of the x and the translation number as people, may be helpful.

$$y = f(x) + a \quad \rightarrow \quad y = f(☺) + ☺$$

<u>Vertical translations are very nice!</u> It is very nice if two people hang out together, but don't get too close. Everyone wants their personal space! So, if the x and the a are not too close, that is very nice and represents a vertical translation.

$$y = f(x + a) \quad \rightarrow \quad y = f(☹+☹)$$

<u>Horizontal translations are horrible!</u> If the x and a get too close, their personal space is reduced and that is horrible! So, when the x and a are both inside the parenthesis, they are too close, so it represents a horrible, horizontal translation.

Note: Function structures, such as square root and absolute value, may be considered a special case of parenthesis. Translations inside these structures also represent horizontal translations (too close).

EQUATIONS TO GRAPHS

The following sections include detailed examples for many function types (parent functions). For each example, the same approach is used so it is easy to see that the same approach works for many function types.

The examples progress from simple to more complex, so reviewing each example, in order, teaches the reader by a series of examples.

In some cases, the given function is rewritten to help identify the translations. For example, the function $y = \sqrt{2x + 6}$ is rewritten as $y = \sqrt{2(x + 3)}$ so the horizontal translations are more easily identified. In this example, there is a horizontal compression by $\frac{1}{2}$ and also a horizontal shift to the left by 3.

Using simple algebra to convert functions to the form:

$y = f(x) = a \cdot f(b(x + c)) + d$ is helpful because the translations are more easily identified, as outlined below.

$y = f(x) = a \cdot f(b(x+c)) + d$	
a	Vertical Stretch by a
b	Horizontal Compression by $\frac{1}{b}$
c	Horizontal Shift to the left by c
d	Vertical Shift up by d

Where: $a \geq 1$, $b \geq 1$, $c \geq 0$, $d \geq 0$

Notice that the vertical translations occur outside of the parenthesis and are fairly straight-forward (very nice). But, the horizontal translations occur within the parenthesis and are the opposite of what you would expect (horrible).

Summary: Vertical translations are outside of the parenthesis and horizontal translations are inside the parenthesis.

Linear Functions

		Example: $y = 3x + 2 = 3(x) + 2$	
Parent Function	$y = x$		
Translations	Vertical		V. Stretch by 3 V. Shift up by 2
	Horizontal		None
Find some (x, y) points. Use one of these two methods.	Use the function	$f(0) = 2$, $f(1) = 5$	
	Use translations to create extended T-table	<table><tr><th>x</th></tr><tr><td>0</td></tr><tr><td>1</td></tr></table>	<table><tr><th>x</th><th>y</th><th>3y + 2</th></tr><tr><td>0</td><td>0</td><td>2</td></tr><tr><td>1</td><td>1</td><td>5</td></tr></table>
Points		$(0, 2) , (1, 5)$	
Asymptotes		None	
Graph of Given Function		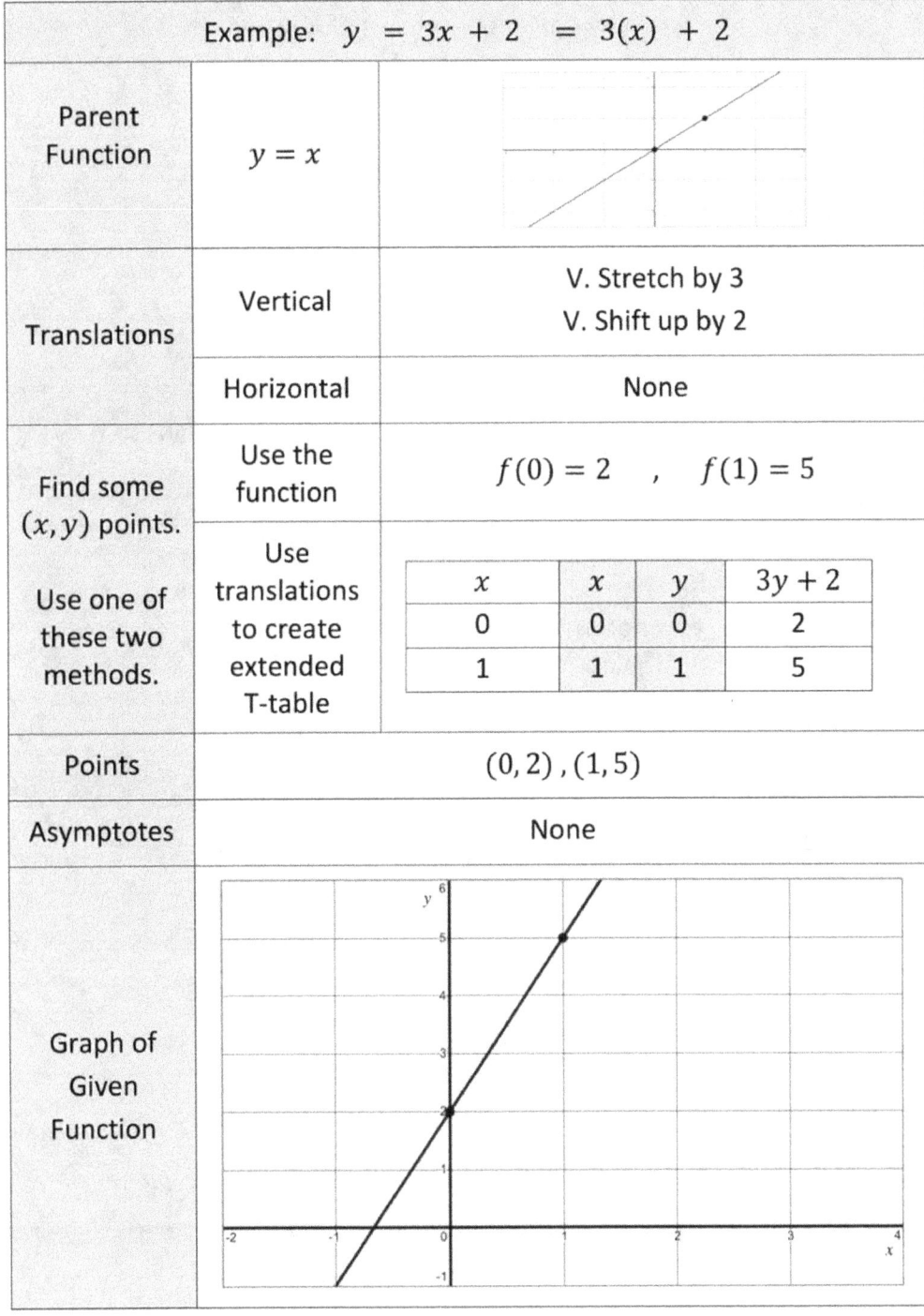	

Quadratic Functions

		Example: $y = (x+1)^2 + 3$	
Parent Function	$y = x^2$		
Translations	Vertical	V. Shift up by 3	
	Horizontal	H. Shift left by 1	
Find some (x, y) points.	Use the function	$f(-1) = 3$, $f(0) = 4$	
Use one of these two methods.	Use translations to create extended T-table	<table><tr><td>$x-1$</td><td>x</td><td>y</td><td>$y+3$</td></tr><tr><td>-1</td><td>0</td><td>0</td><td>3</td></tr><tr><td>0</td><td>1</td><td>1</td><td>4</td></tr></table>	
Points		$(-1, 3), (0, 4)$	
Asymptotes		None	
Graph of Given Function	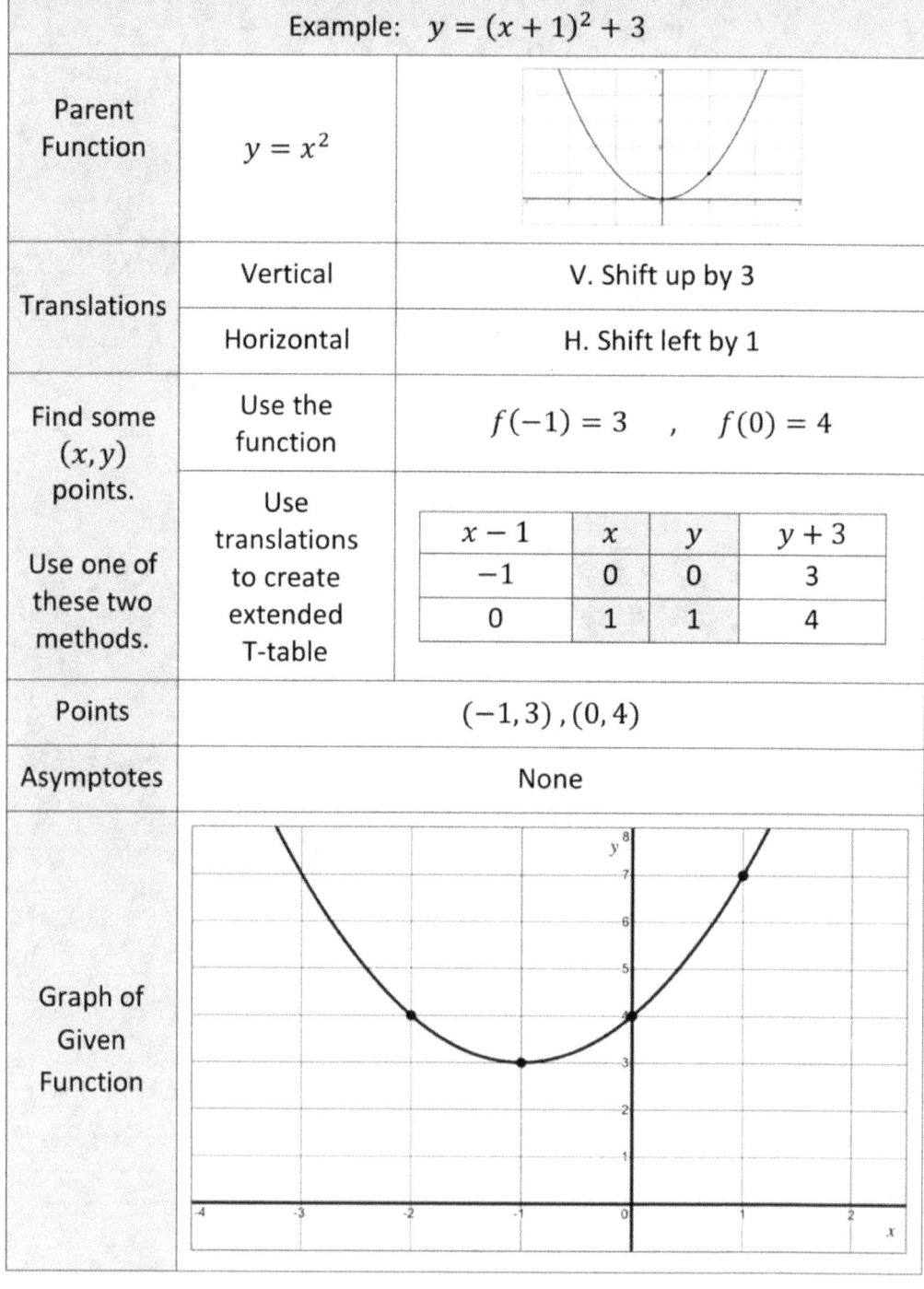		

Cubic Functions

		Example: $y = (x-1)^3 - 4$				
Parent Function	$y = x^3$	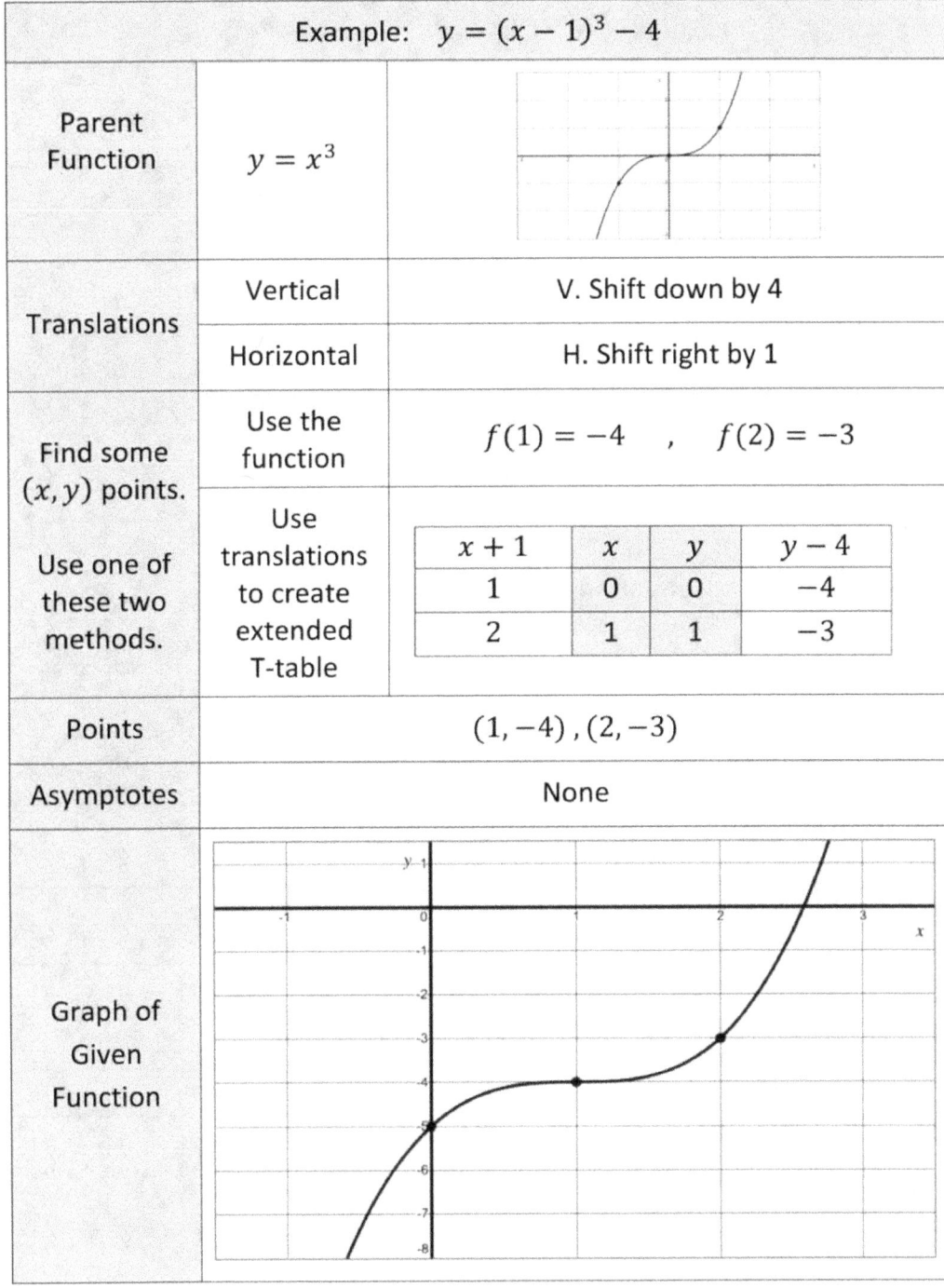				
Translations	Vertical	V. Shift down by 4				
	Horizontal	H. Shift right by 1				
Find some (x, y) points. Use one of these two methods.	Use the function	$f(1) = -4$, $f(2) = -3$				
	Use translations to create extended T-table	$x+1$	x	y	$y-4$	
		1	0	0	-4	
		2	1	1	-3	
Points		$(1, -4), (2, -3)$				
Asymptotes		None				
Graph of Given Function						

Exponential Functions

		Example: $y = -3^x + 2 = -(3^x) + 2$	
Parent Function	$y = 3^x$		
Translations	Vertical	V. Rotation V. Shift up by 2	
	Horizontal	None	
Find some (x,y) points.	Use the function	$f(0) = 1$, $f(1) = -1$	
Use one of these two methods.	Use translations to create extended T-table	(see table below)	
Points		$(0, 1), (1, -1)$	
Asymptotes		$y = 2$	
Graph of Given Function			

Extended T-table:

x	x	y	$-y + 2$
0	0	1	1
1	1	3	-1

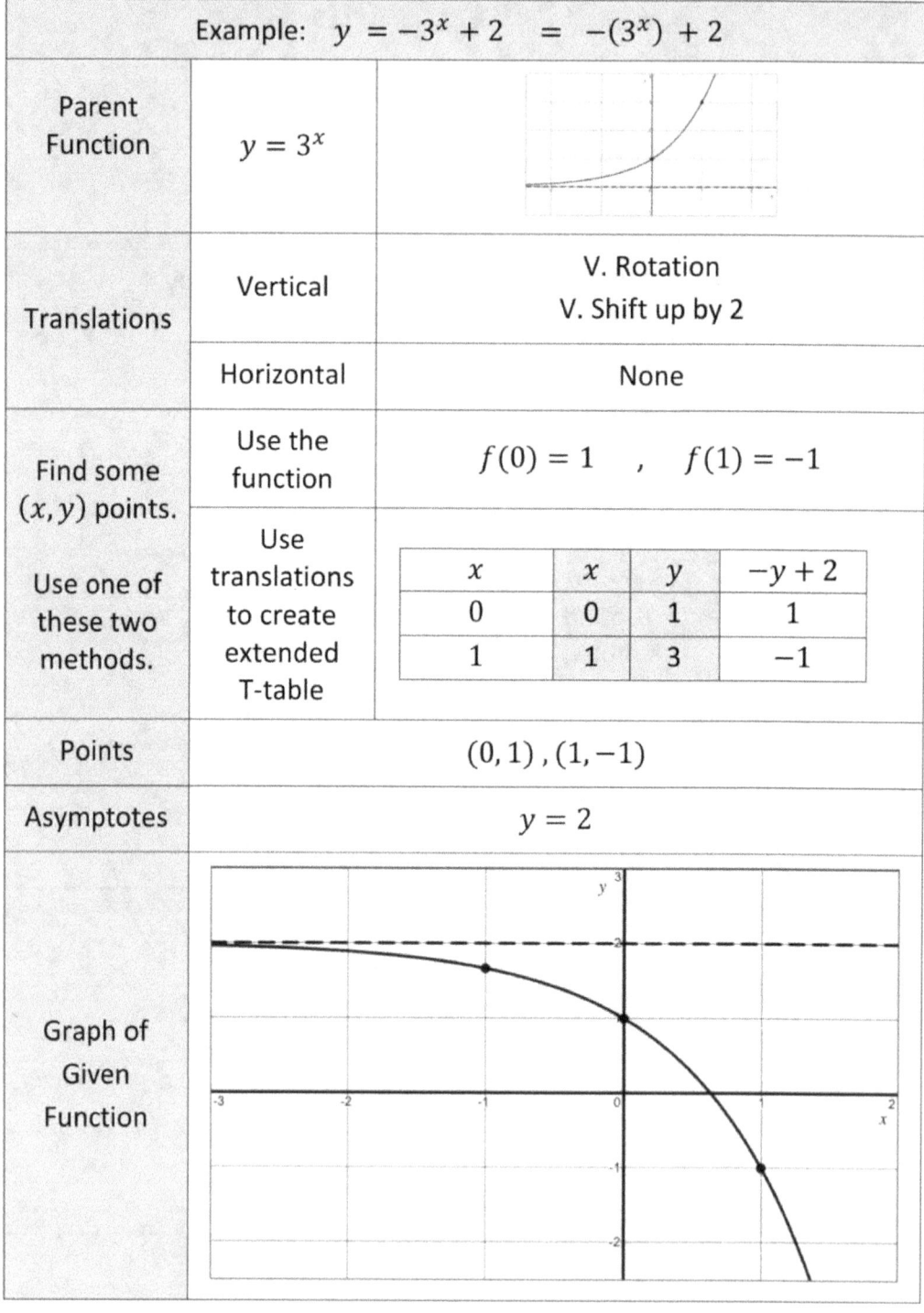

Absolute Value Functions

		Example: $y = -2|x + 3| + 4$	
Parent Function	$y = \|x\|$		
Translations	Vertical		V. Rotation V. Stretch by 2 V. Shift up by 4
	Horizontal		H. Shift left by 3
Find some (x, y) points. Use one of these two methods.	Use the function	$f(-3) = 4$, $f(-2) = 2$	
	Use translations to create extended T-table	<table><tr><td>$x - 3$</td><td>x</td><td>y</td><td>$-2y + 4$</td></tr><tr><td>-3</td><td>0</td><td>0</td><td>4</td></tr><tr><td>-2</td><td>1</td><td>1</td><td>2</td></tr></table>	
Points		$(-3, 4), (-2, 2)$	
Asymptotes		None	
Graph of Given Function		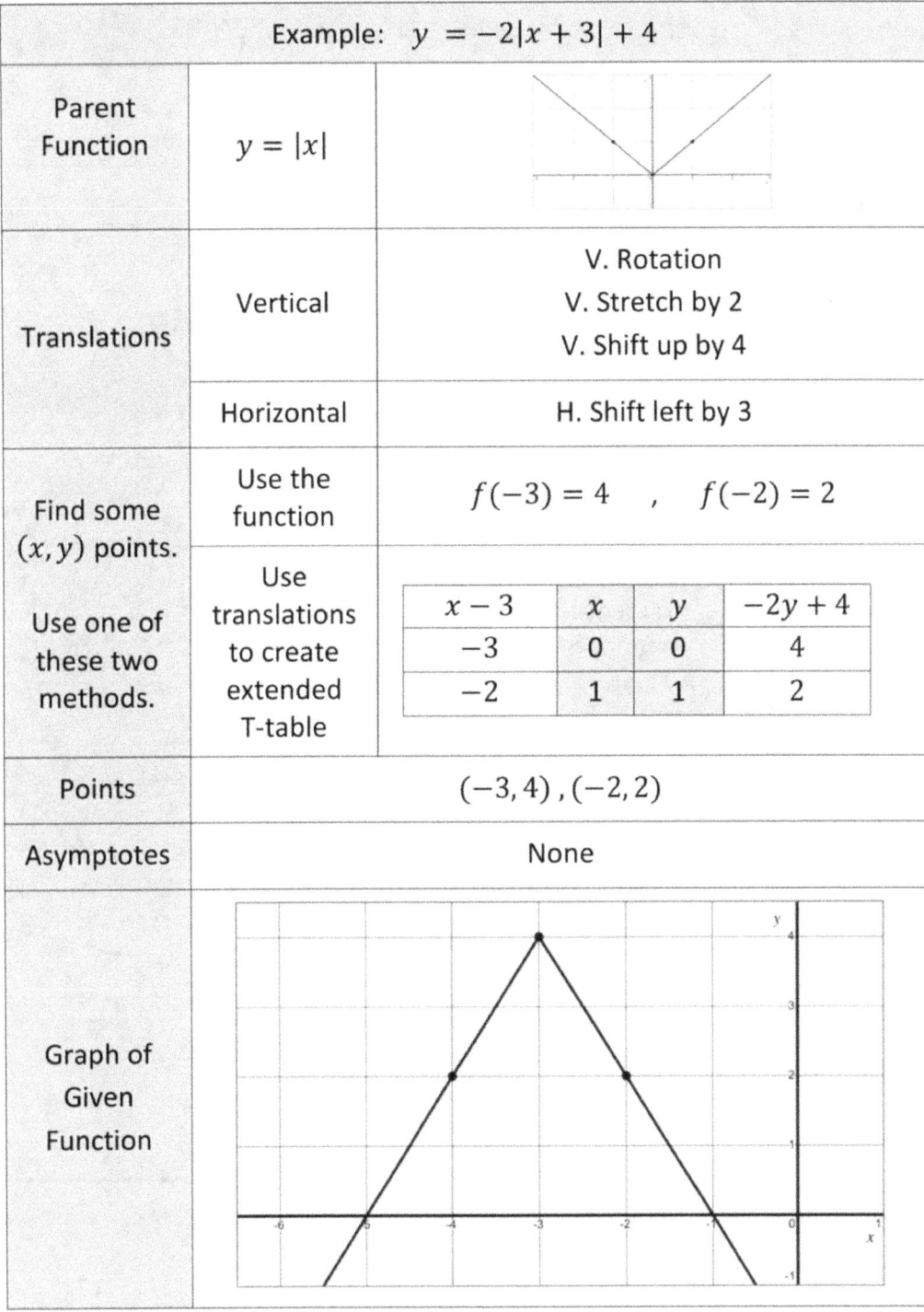	

Square Root Functions

		Example: $y = \sqrt{2x+6} = \sqrt{2(x+3)}$	
Parent Function	$y = \sqrt{x}$		
Translations	Vertical	None	
	Horizontal	H. Compression by $\frac{1}{2}$ H. Shift left by 3	
Find some (x,y) points. Use one of these two methods.	Use the function	$f(-3) = 0$, $f(-1) = 2$	
	Use translations to create extended T-table	$\begin{array}{\|c\|c\|c\|} \hline \frac{x}{2}-3 & x & y \\ \hline -3 & 0 & 0 \\ \hline -1 & 4 & 2 \\ \hline \end{array}$ ... $\begin{array}{\|c\|} \hline y \\ \hline 0 \\ \hline 2 \\ \hline \end{array}$	
Points		$(-3, 0), (-1, 2)$	
Asymptotes		None	
Graph of Given Function	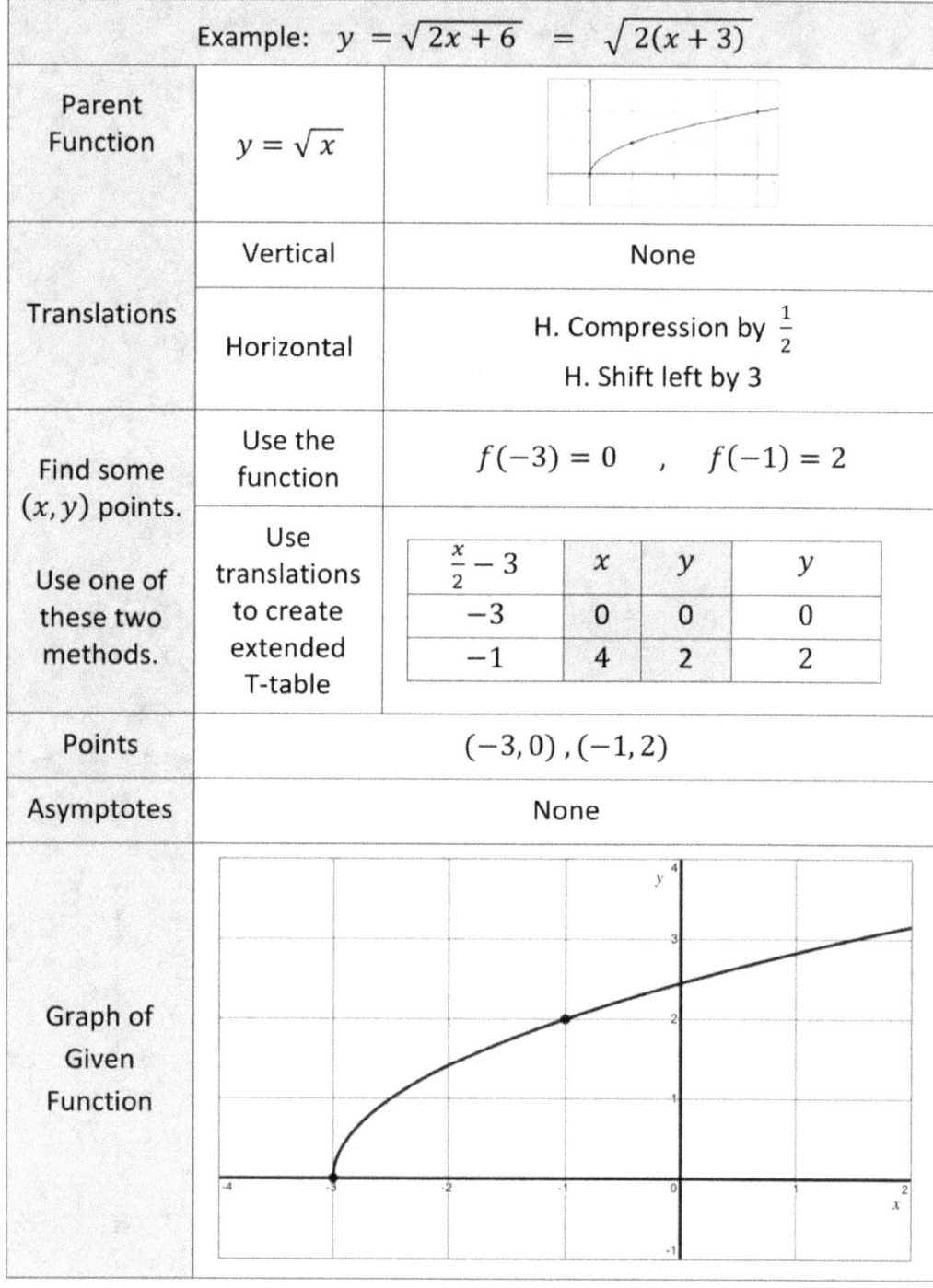		

Logarithmic Functions

		Example: $y = \log(-x + 3) + 2 = \log(-(x - 3)) + 2$					
Parent Function	$y = \log(x)$	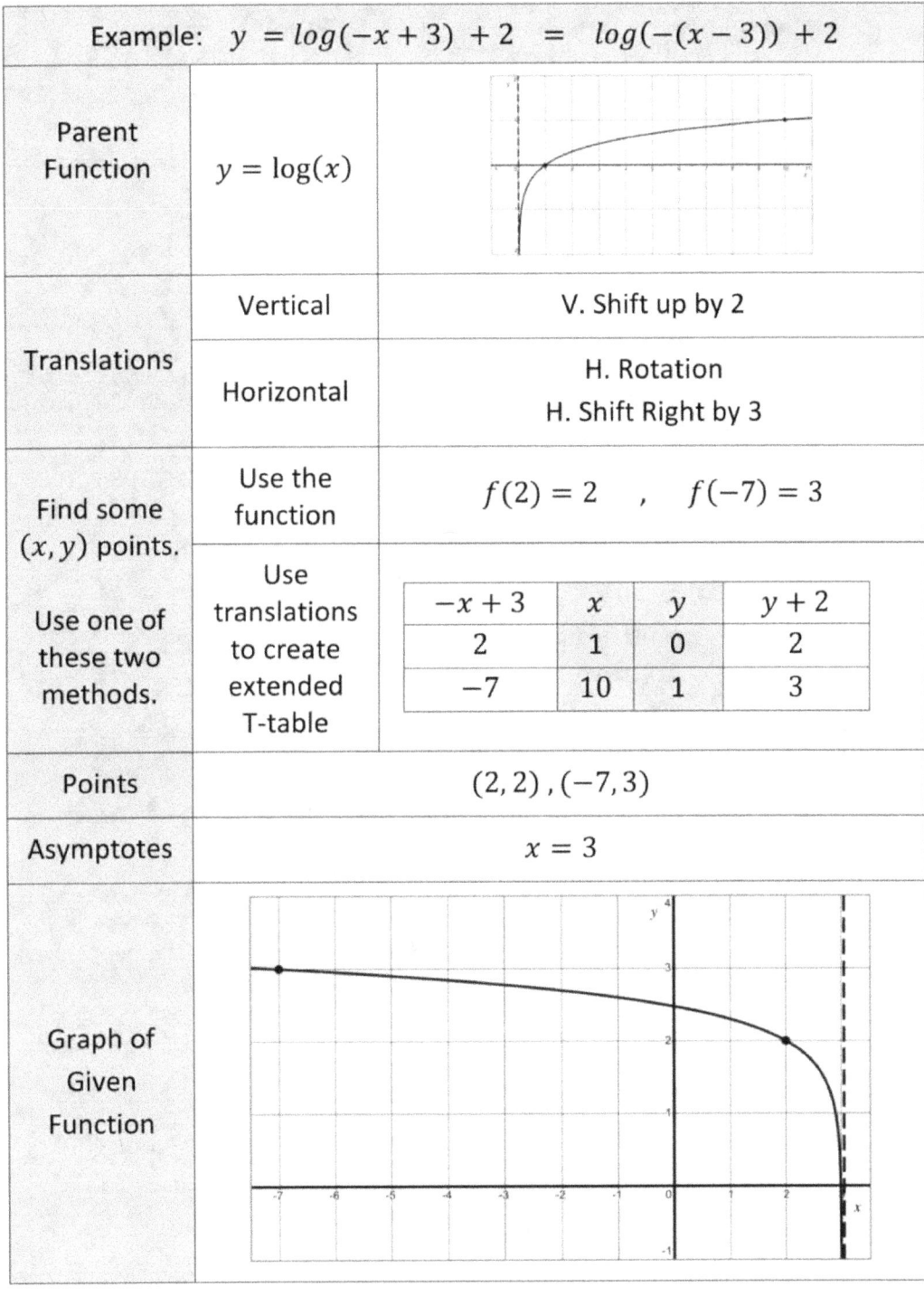					
Translations	Vertical	V. Shift up by 2					
	Horizontal	H. Rotation H. Shift Right by 3					
Find some (x, y) points. Use one of these two methods.	Use the function	$f(2) = 2$, $f(-7) = 3$					
	Use translations to create extended T-table	$-x + 3$	x	y	$y + 2$		
		2	1	0	2		
		-7	10	1	3		
Points		$(2, 2), (-7, 3)$					
Asymptotes		$x = 3$					
Graph of Given Function							

Rational Functions (part 1)

Example: $y = -\frac{3}{x} + 2 = -3\left(\frac{1}{x}\right) + 2$

Parent Function	$y = \frac{1}{x}$	
Translations	Vertical	V. Rotation over x-axis V. Stretch by 3 V. Shift up by 2
	Horizontal	None
Find some (x, y) points. Use one of these two methods.	Use the function	$f(1) = 5$, $f(-1) = -1$
	Use translations to create extended T-table	
Points		$(1, -1), (-1, 5)$
Asymptotes		$x = 0$ and $y = 2$
Graph of Given Function		

Extended T-table:

x	x	y	$-3y + 2$
1	1	1	-1
-1	-1	-1	5

Rational Functions (part 2)

		Example: $y = -\dfrac{1}{(x+3)^2}$			
Parent Function	$y = \dfrac{1}{x^2}$	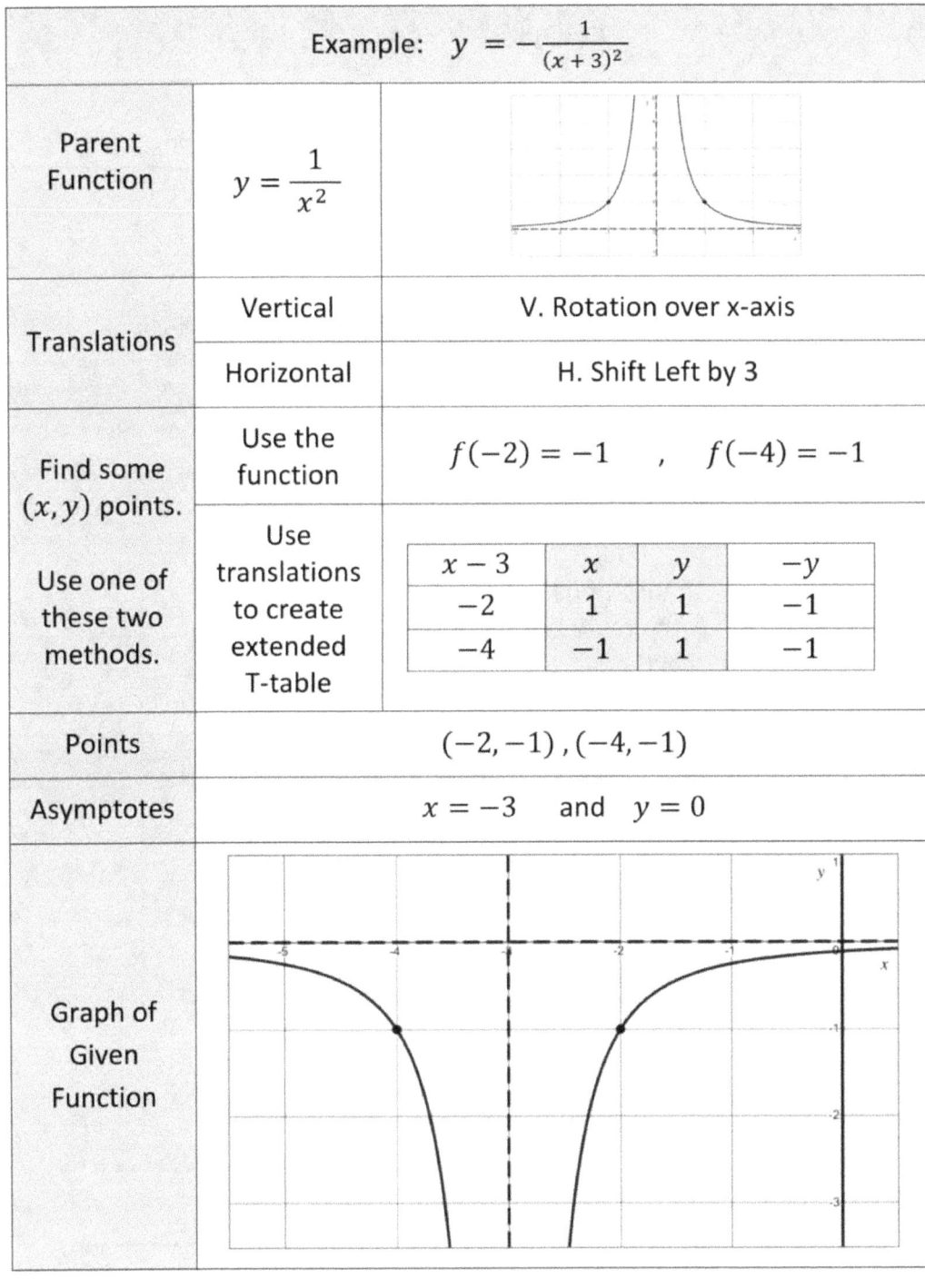			
Translations	Vertical	V. Rotation over x-axis			
	Horizontal	H. Shift Left by 3			
Find some (x, y) points.	Use the function	$f(-2) = -1$, $f(-4) = -1$			
Use one of these two methods.	Use translations to create extended T-table	$x-3$	x	y	$-y$
		-2	1	1	-1
		-4	-1	1	-1
Points		$(-2, -1), (-4, -1)$			
Asymptotes		$x = -3$ and $y = 0$			
Graph of Given Function					

Step Functions (Greatest Integer <= x)

		Example: $y = -\lfloor x+3 \rfloor$				
Parent Function	$y = \lfloor x \rfloor$					
Translations	Vertical	V. Rotation over x-axis				
	Horizontal	H. Shift Left by 3				
Find some (x,y) points. Use one of these two methods.	Use the function	$f(-3) = 0$, $f(-2) = -1$				
	Use translations to create extended T-table	$x - 3$	x	y	$-y$	
		-3	0	0	0	
		-2	1	1	-1	
Points		$(-3, 0), (-2, -1)$				
Asymptotes		None				
Graph of Given Function						

	Example: $y = \left\lfloor -\frac{x}{2} - 1 \right\rfloor = \left\lfloor -\frac{1}{2}(x+2) \right\rfloor$						
Parent Function	$y = \lfloor x \rfloor$						
Translations	Vertical		None				
	Horizontal		H. Rotation over y-axis H. Stretch by 2 H. Shift Left by 2				
Find some (x, y) points. Use one of these two methods.	Use the function	$f(-2) = 0$, $f(-4) = 1$					
	Use translations to create extended T-table	$-2x - 2$	x	y		y	
		-2	0	0		0	
		-4	1	1		1	
Points	$(-2, 0)$, $(-4, 1)$						
Asymptotes	None						
Graph of Given Function							

Sine Functions

Example:	$y = 3\sin(4x + \pi) + 5 = 3\sin\left(4\left(x + \frac{\pi}{4}\right)\right) + 5$		
Parent Function	$y = \sin(x)$ Period $= 2\pi$		
Translations	Vertical	V. Stretch by 3 → Amplitude = 3 V. Shift up by 5	
	Horizontal	H. Comp. by $\frac{1}{4}$ → Period $= \frac{2\pi}{4} = \frac{\pi}{2}$ H. Shift Left by $\frac{\pi}{4}$ → Start at $-\frac{\pi}{4}$	
Find some (x, y) points. Use one of these two methods.	Use the function	$f\left(-\frac{\pi}{4}\right) = 5$, $f\left(\frac{3\pi}{4}\right) = 5$	
	Use translations to create extended T-table	$\left(\frac{1}{4}\right)x - \frac{\pi}{4}$ | x | y | $y + 5$ $-\frac{\pi}{4}$ | 0 | 0 | 5 $\frac{3\pi}{4}$ | 2π | 0 | 5	
Points	$\left(-\frac{\pi}{4}, 5\right), \left(\frac{3\pi}{4}, 5\right)$		
Graph of Given Function	(graph showing sine curve with Period marked, x-axis from $-\pi/2$ to $\pi/2$, y-axis up to 8)		

Example:	$y = 3\sin(4x + 180°) + 5 = 3\sin(4(x + 45°)) + 5$		
Parent Function	$y = \sin(x)$ Period = 360°		
Translations		Vertical	V. Stretch by 3 → Amplitude = 3 V. Shift up by 5
		Horizontal	H. Comp. by $\frac{1}{4}$ → Period = $\frac{360}{4}$ = 90° H. Shift Left by 45° → Start at −45°
Find some (x, y) points. Use one of these two methods.		Use the function	$f(-45°) = 5$, $f(45°) = 5$
		Use translations to create extended T-table	<table><tr><td>$\left(\frac{1}{4}\right)x - 45$</td><td>$x$</td><td>$y$</td><td>$y + 5$</td></tr><tr><td>−45°</td><td>0</td><td>0</td><td>5</td></tr><tr><td>45°</td><td>360</td><td>0</td><td>5</td></tr></table>
Points	$(-45°, 5), (45°, 5)$		
Graph of Given Function			

Cosine Functions

Example:	$y = -3\cos\left(\frac{x}{2} - \frac{\pi}{3}\right) + 5 = -3\cos\left(\frac{1}{2}\left(x - \frac{2\pi}{3}\right)\right) + 5$	
Parent Function	$y = \cos(x)$ Period $= 2\pi$	
Translations	Vertical	V. Rotation over x-axis V. Stretch by 3 → Amplitude = 3 V. Shift up by 5
	Horizontal	H. Stretch by 2 → Period $= 2(2\pi) = 4\pi$ H. Shift Right by $\frac{2\pi}{3}$ → Start at $\frac{2\pi}{3}$
Find some (x, y) points. Use one of these two methods.	Use the function	$f\left(\frac{2\pi}{3}\right) = 2$, $f\left(\frac{14\pi}{3}\right) = 2$
	Use translations to create extended T-table	$\begin{array}{\|c\|c\|c\|c\|} \hline 2x + \frac{2\pi}{3} & x & y & -3y+5 \\ \hline \frac{2\pi}{3} & 0 & 1 & 2 \\ \hline \frac{14\pi}{3} & 2\pi & 1 & 2 \\ \hline \end{array}$
Points		$\left(\frac{2\pi}{3}, 2\right)$, $\left(\frac{14\pi}{3}, 2\right)$
Graph of Given Function		(graph of cosine function with period marked from $2\pi/3$ to $14\pi/3$; x-axis marked at $2\pi/3$, $4\pi/3$, 2π, $8\pi/3$, $10\pi/3$, 4π, $14\pi/3$)

Example:	$y = -4\cos(3x - 2) + 2 = -4\cos\left(3\left(x - \frac{2}{3}\right)\right) + 2$		
Parent Function	$y = \cos(x)$ Period $= 2\pi$		
Translations Assume units are radians	Vertical	V. Rotation over x-axis V. Stretch by 4 → Amplitude = 4 V. Shift up by 2	
	Horizontal	H. Comp. by 3 → Period $= \frac{2\pi}{3} = 2.09$ H. Shift Right by $\frac{2}{3}$ → Start at $\frac{2}{3}$	
Find some (x, y) points.	Use translations to create extended T-table	$\frac{1}{3}(x) + \frac{2}{3}$ x y $-4y + 2$ $\frac{2}{3} = .7$ 0 1 -2 $\frac{\pi}{3} + \frac{2}{3} = 1.7$ π -1 6 $\frac{2\pi}{3} + \frac{2}{3} = 2.8$ 2π 1 -2	
Points	$(0.7, -2)$, $(1.7, 6)$, $(2.8, -2)$		
Graph of Given Function Plot points first			

Tangent Functions

		Example: $y = 2\tan\left(x + \frac{\pi}{4}\right) + 3$			
Parent Function	$y = \tan(x)$ Period $= \pi$				
Translations	Vertical	V. Stretch by 2 , V. Shift up by 3			
	Horizontal	H. Shift Left by $\frac{\pi}{4}$ → Start at $-\frac{\pi}{4}$			
Find some (x, y) points. Use one of these two methods.	Use the function	$f\left(-\frac{\pi}{4}\right) = 3$, $f(0) = 7$			
	Use translations to create extended T-table	$x - \frac{\pi}{4}$	x	y	$2y + 3$
		$-\frac{\pi}{4}$	0	0	3
		0	$\frac{\pi}{4}$	1	7
Points		$\left(-\frac{\pi}{4}, 3\right)$, $(0, 7)$			
Graph of Given Function	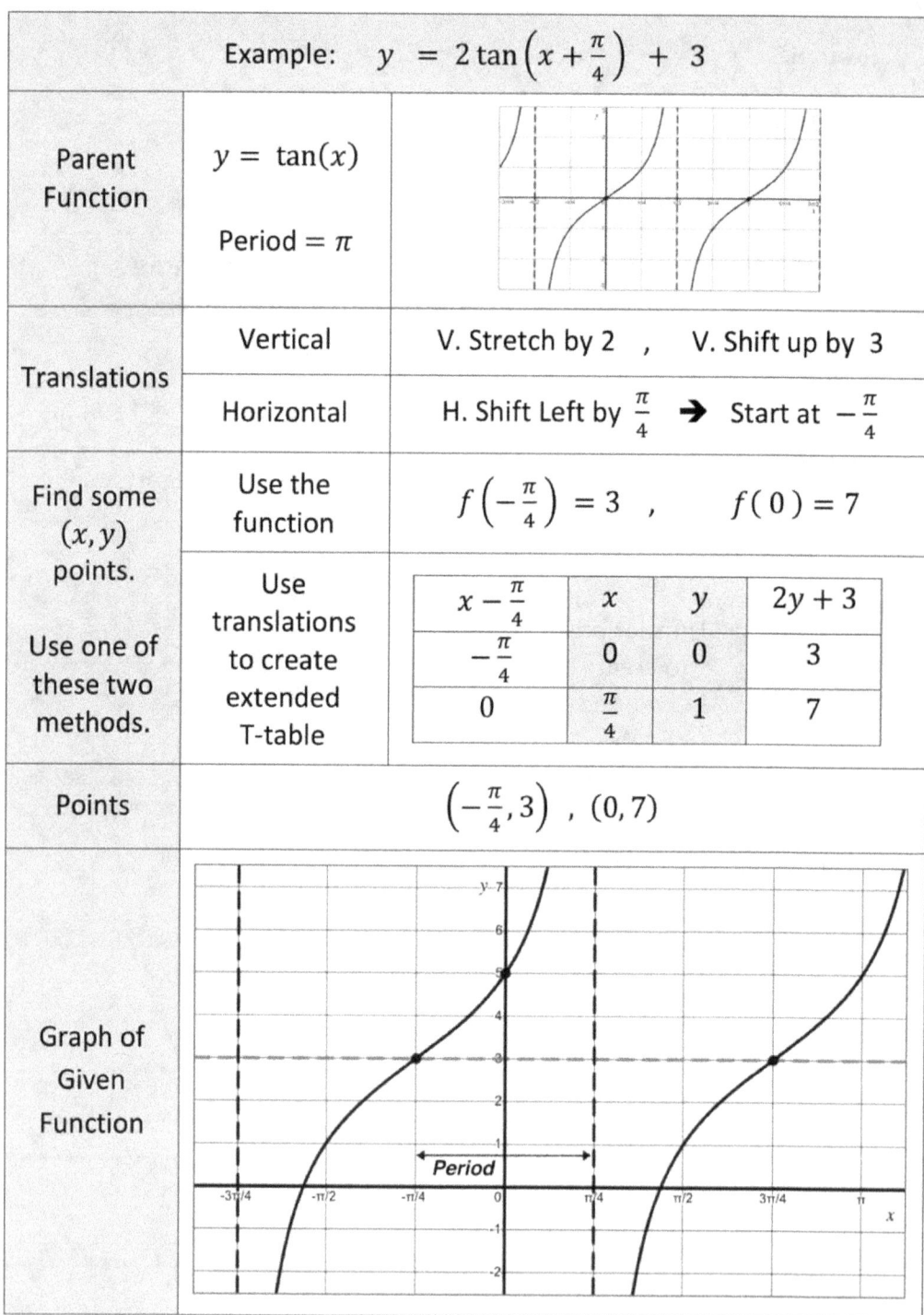				

GRAPHS TO EQUATIONS

The following sections include detailed examples for many function types. For each example, the same process is used.

As noted in the previous section, using simple algebra to convert functions to the form: $y = f(x) = a \cdot f(b(x+c)) + d$ is helpful because the translations (vertical and horizontal) are more easily identified, as outlined in the following table.

$y = f(x) = a \cdot f(b(x+c)) + d$	
a	Vertical Stretch by a
b	Horizontal Compression by $\frac{1}{b}$
c	Horizontal Shift to the left by c
d	Vertical Shift up by d

Where: $a \geq 1$, $b \geq 1$, $c \geq 0$, $d \geq 0$

Remember: Vertical translations are outside of the parenthesis and horizontal translations are inside the parenthesis.

Linear Functions

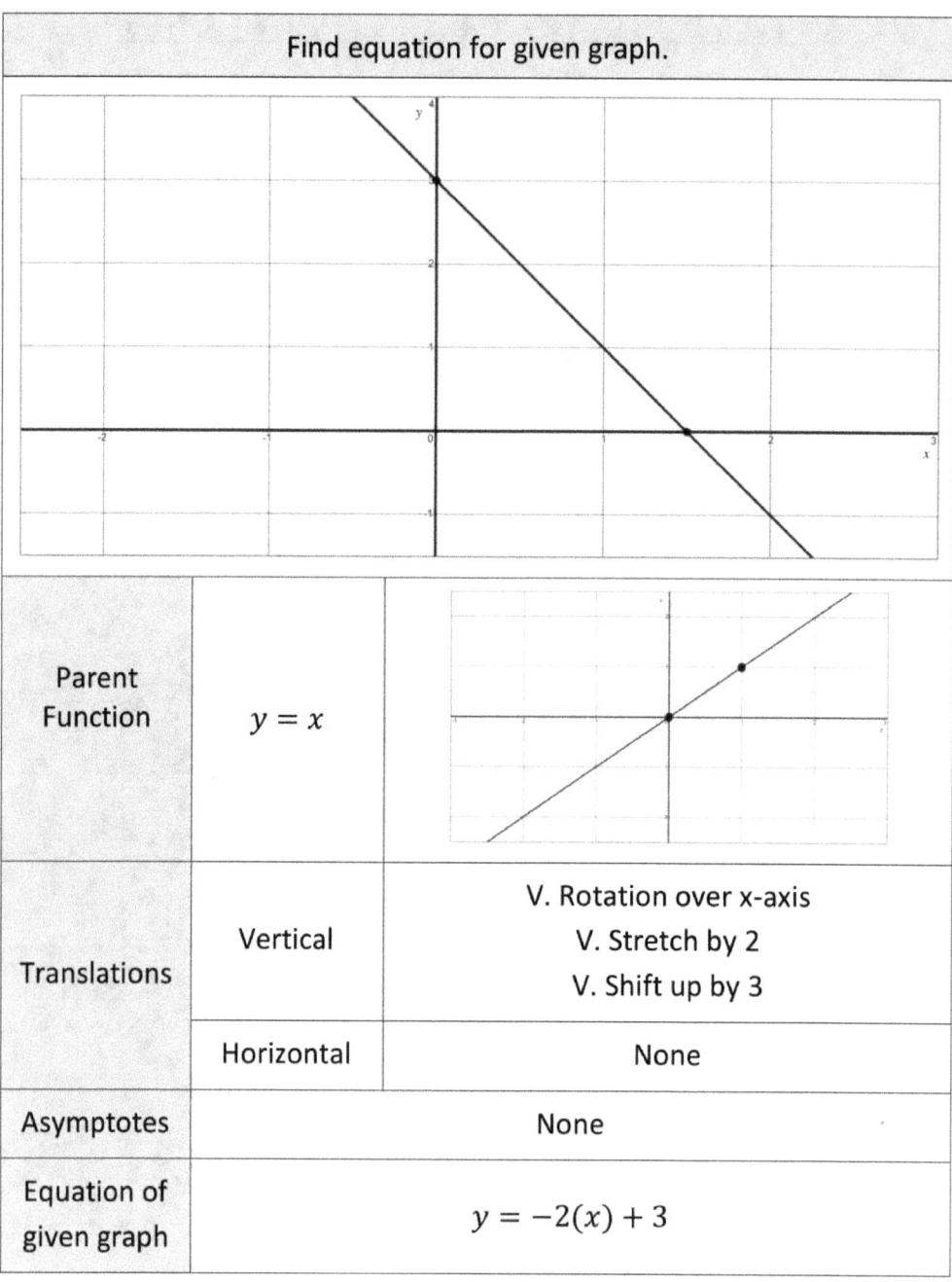

	Find equation for given graph.	
Parent Function	$y = x$	
Translations	Vertical	V. Rotation over x-axis V. Stretch by 2 V. Shift up by 3
	Horizontal	None
Asymptotes		None
Equation of given graph		$y = -2(x) + 3$

Quadratic Functions

		Find equation for given graph.	
		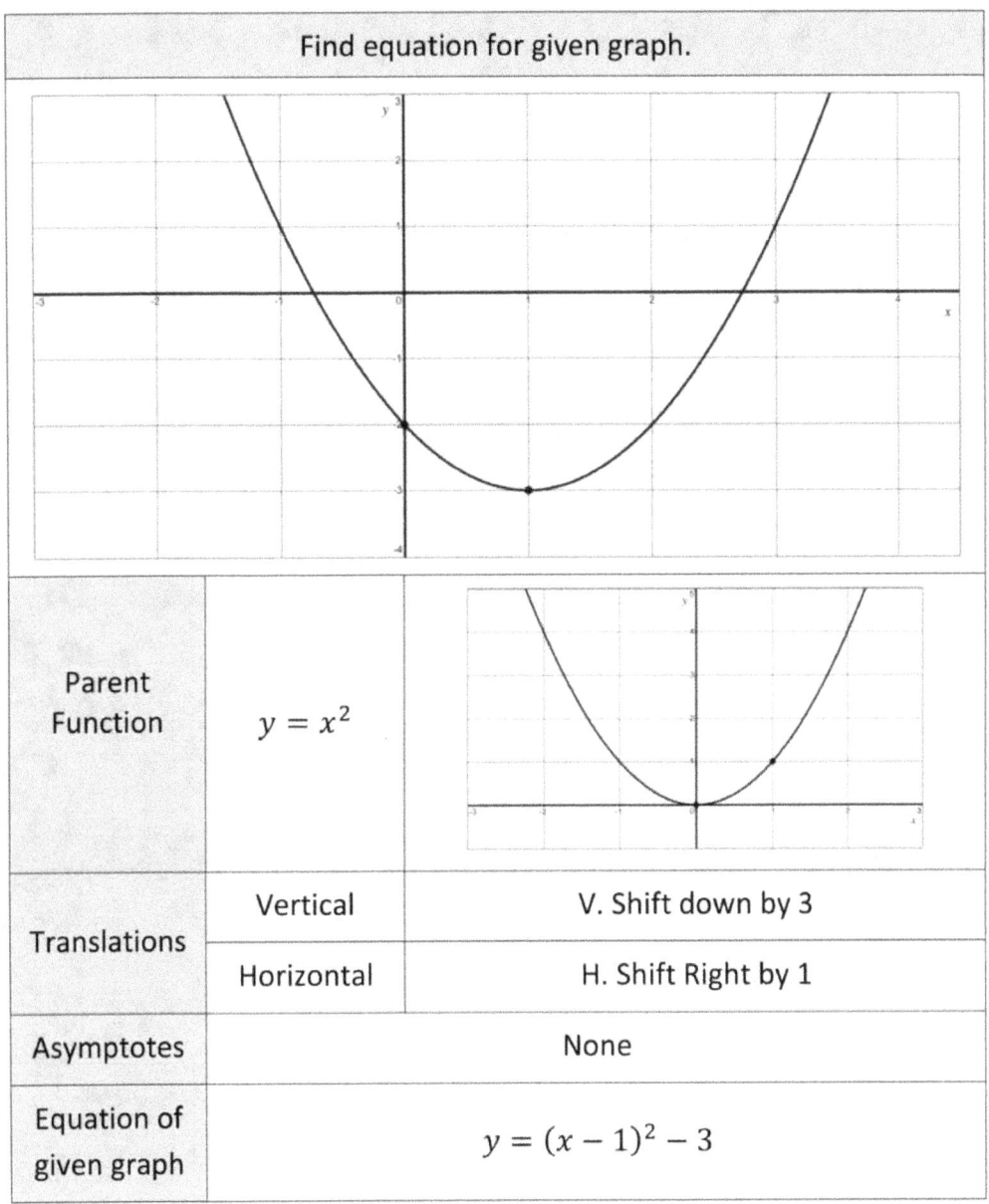	
Parent Function	$y = x^2$		
Translations	Vertical	V. Shift down by 3	
	Horizontal	H. Shift Right by 1	
Asymptotes		None	
Equation of given graph		$y = (x-1)^2 - 3$	

Cubic Functions

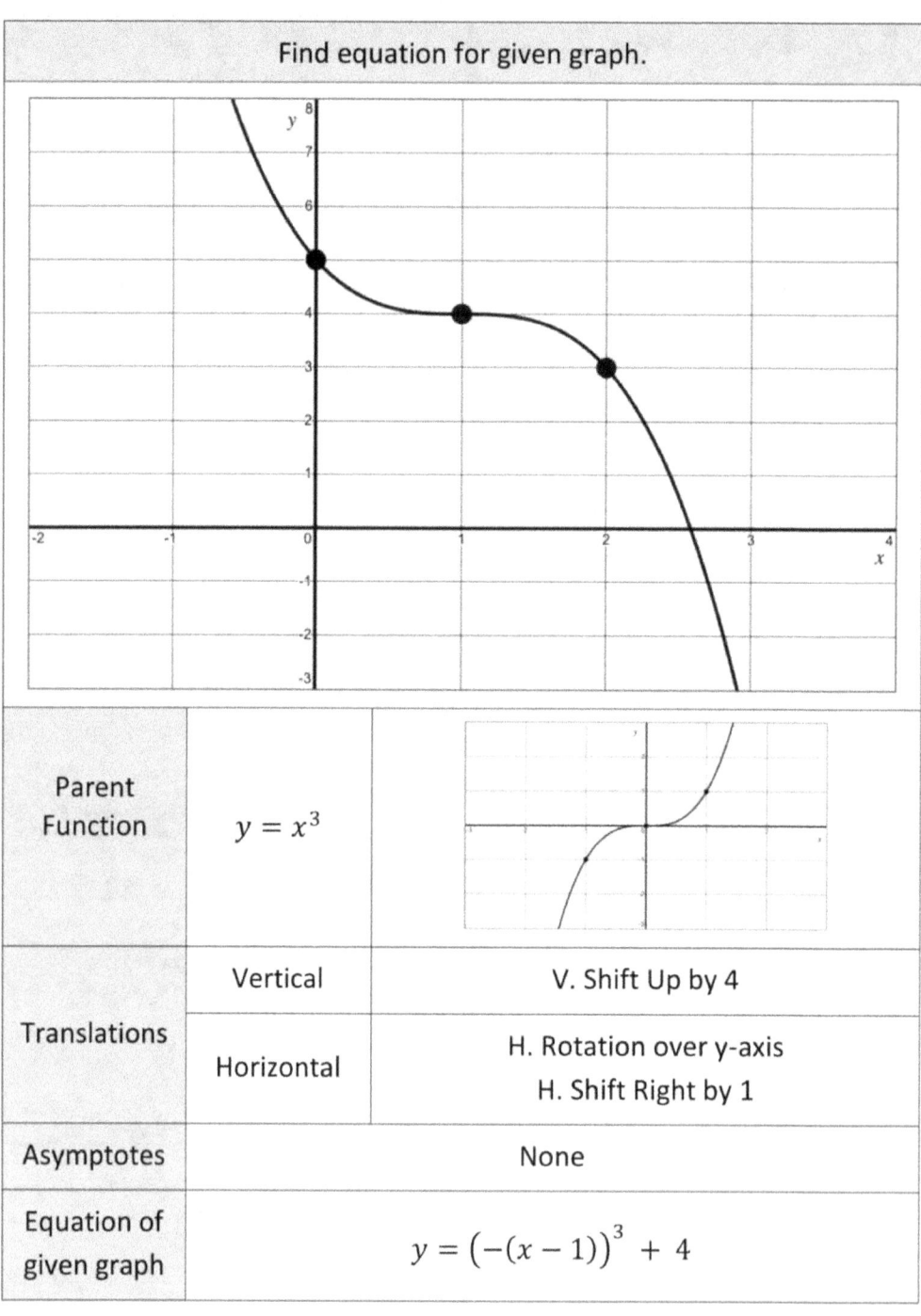

	Find equation for given graph.	
Parent Function	$y = x^3$	
Translations	Vertical	V. Shift Up by 4
	Horizontal	H. Rotation over y-axis H. Shift Right by 1
Asymptotes		None
Equation of given graph		$y = (-(x-1))^3 + 4$

Find equation for given graph.		
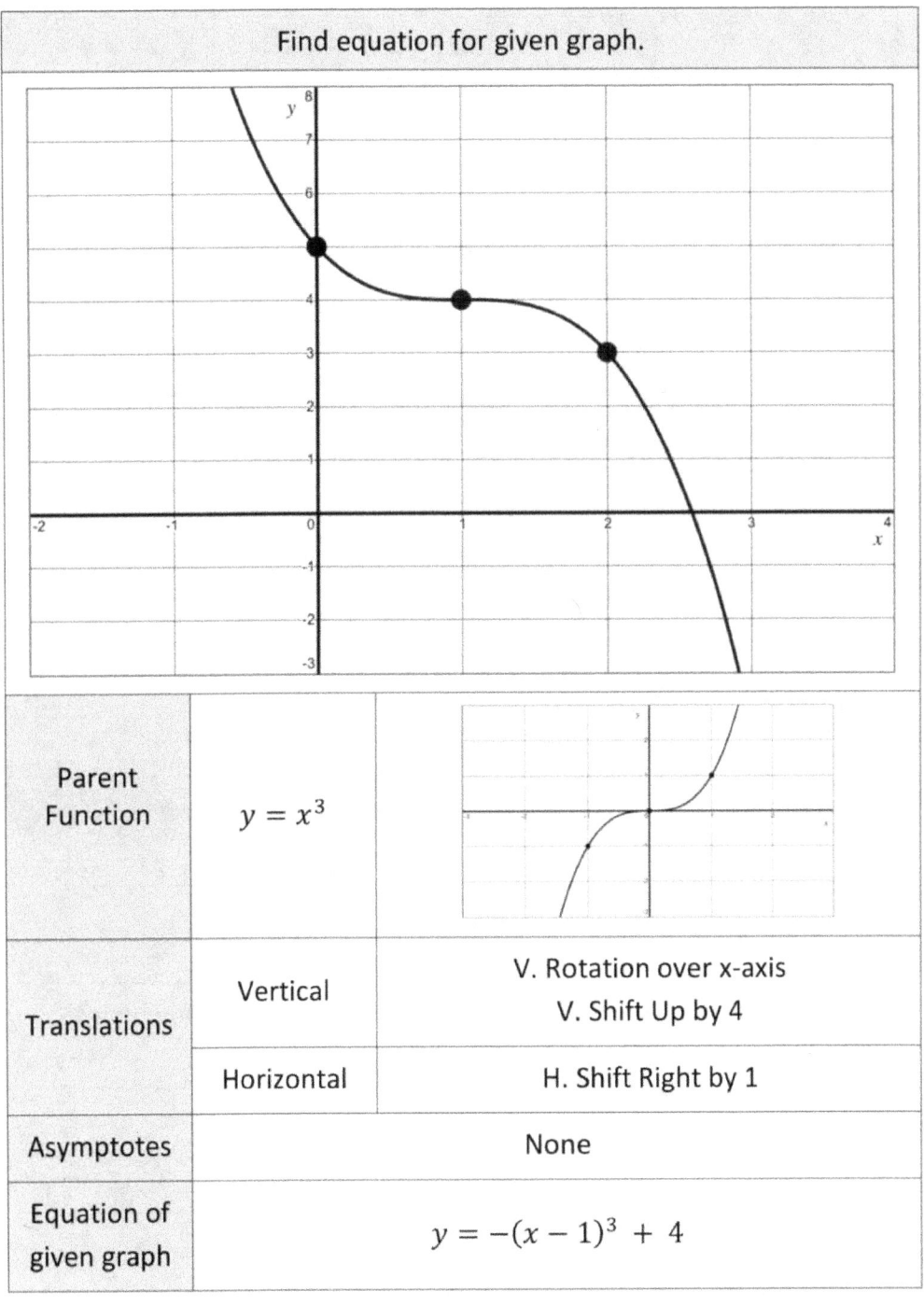		
Parent Function	$y = x^3$	
Translations	Vertical	V. Rotation over x-axis V. Shift Up by 4
	Horizontal	H. Shift Right by 1
Asymptotes		None
Equation of given graph		$y = -(x-1)^3 + 4$

Exponential Functions

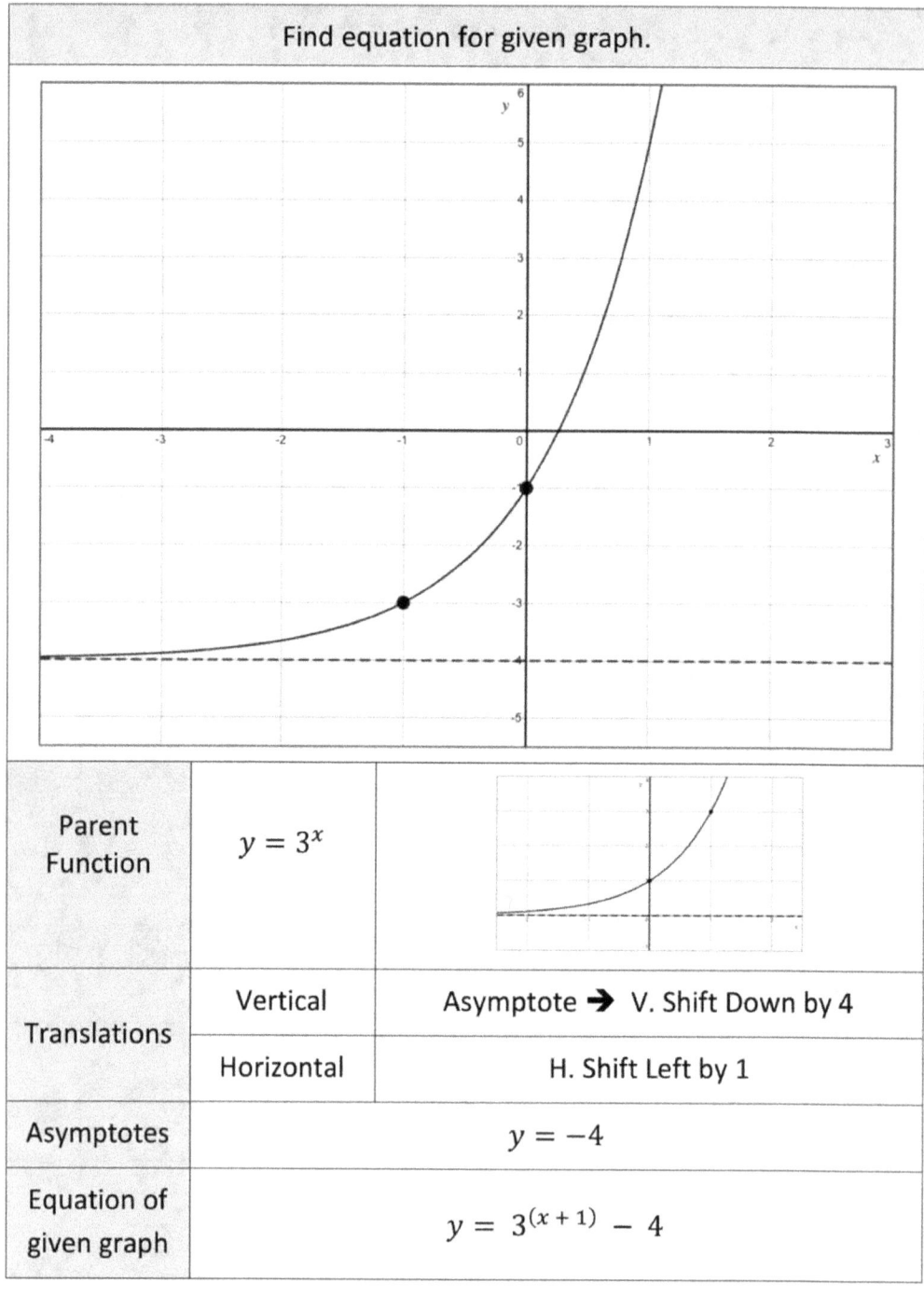

	Find equation for given graph.	
Parent Function	$y = 3^x$	
Translations	Vertical	Asymptote → V. Shift Down by 4
	Horizontal	H. Shift Left by 1
Asymptotes		$y = -4$
Equation of given graph		$y = 3^{(x+1)} - 4$

Absolute Value Functions

	Find equation for given graph.	
	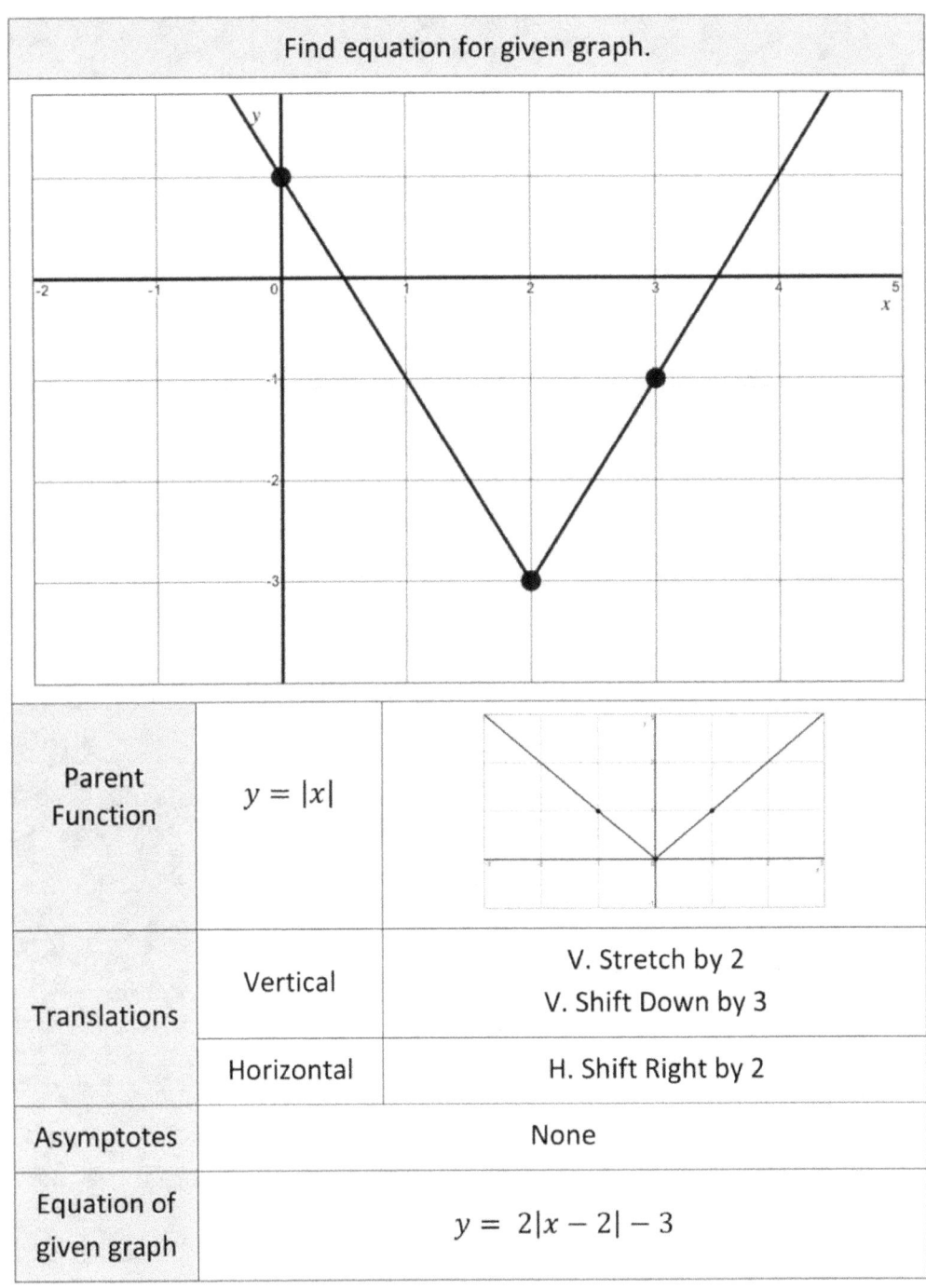	
Parent Function	$y = \|x\|$	
Translations	Vertical	V. Stretch by 2 V. Shift Down by 3
	Horizontal	H. Shift Right by 2
Asymptotes		None
Equation of given graph		$y = 2\|x - 2\| - 3$

Square Root Functions

Find equation for given graph.		
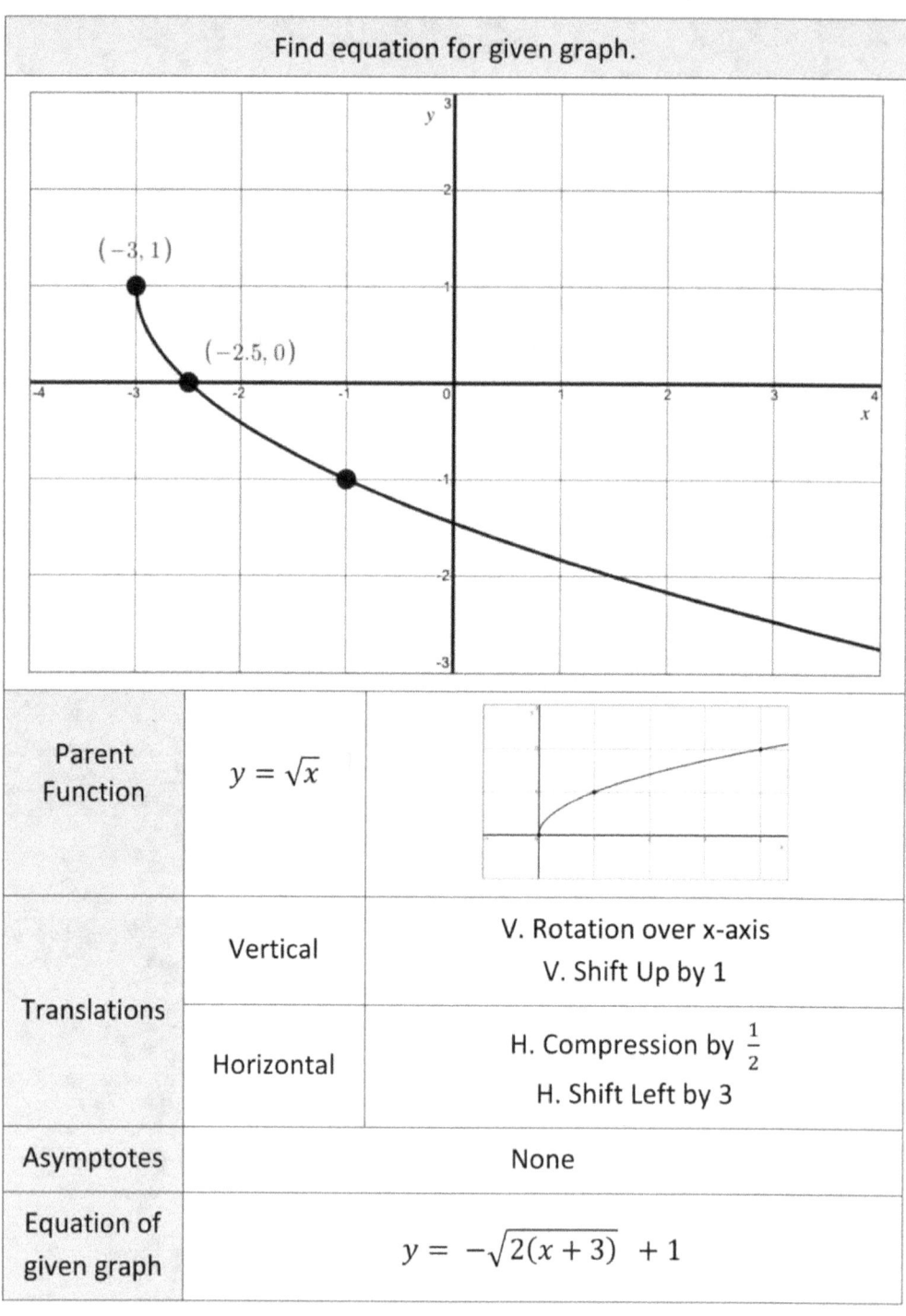		
Parent Function	$y = \sqrt{x}$	
Translations	Vertical	V. Rotation over x-axis V. Shift Up by 1
	Horizontal	H. Compression by $\frac{1}{2}$ H. Shift Left by 3
Asymptotes		None
Equation of given graph		$y = -\sqrt{2(x+3)} + 1$

Logarithmic Functions

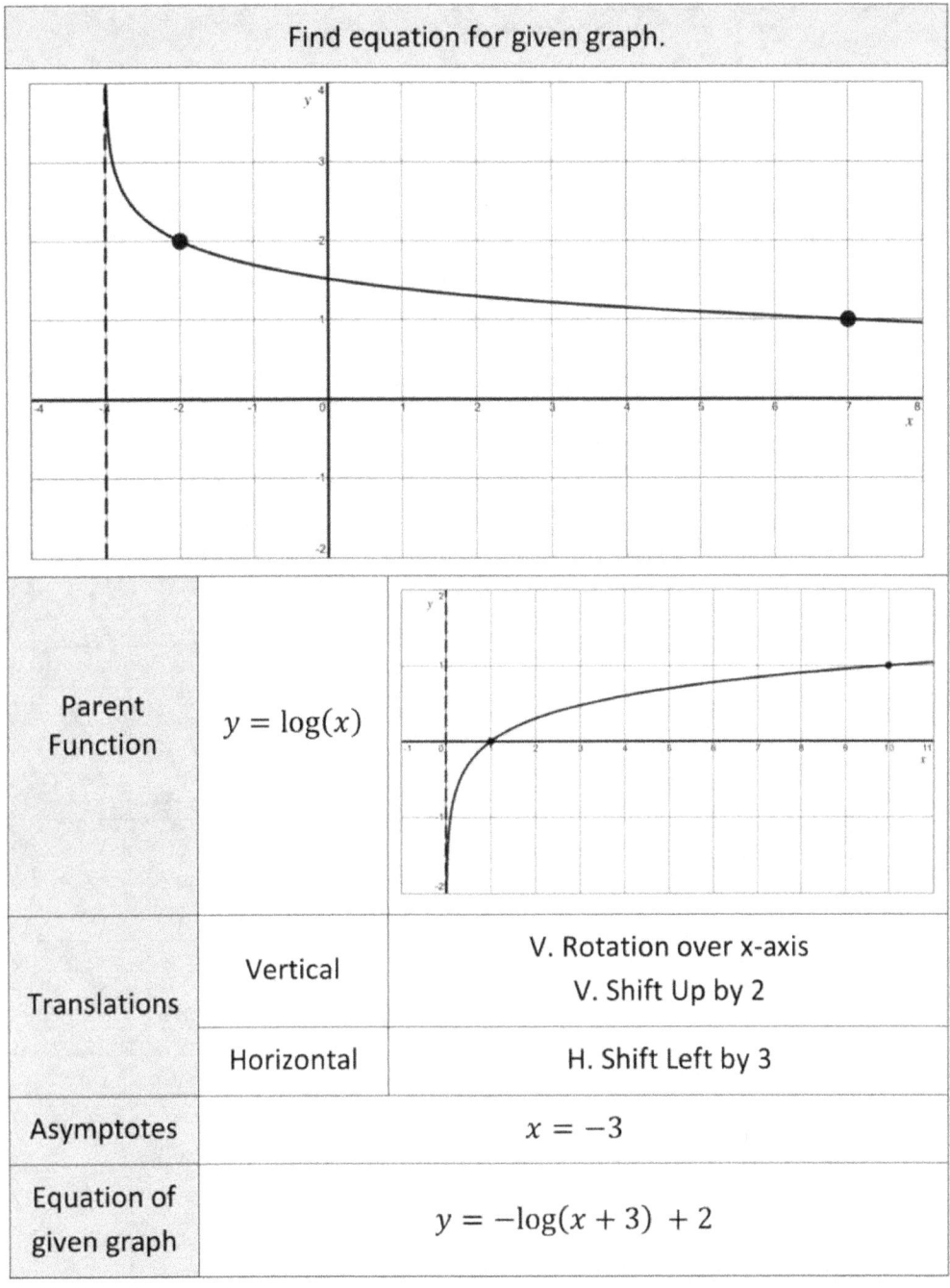

		Find equation for given graph.
Parent Function	$y = \log(x)$	
Translations	Vertical	V. Rotation over x-axis V. Shift Up by 2
	Horizontal	H. Shift Left by 3
Asymptotes		$x = -3$
Equation of given graph		$y = -\log(x+3) + 2$

Rational Functions (part 1)

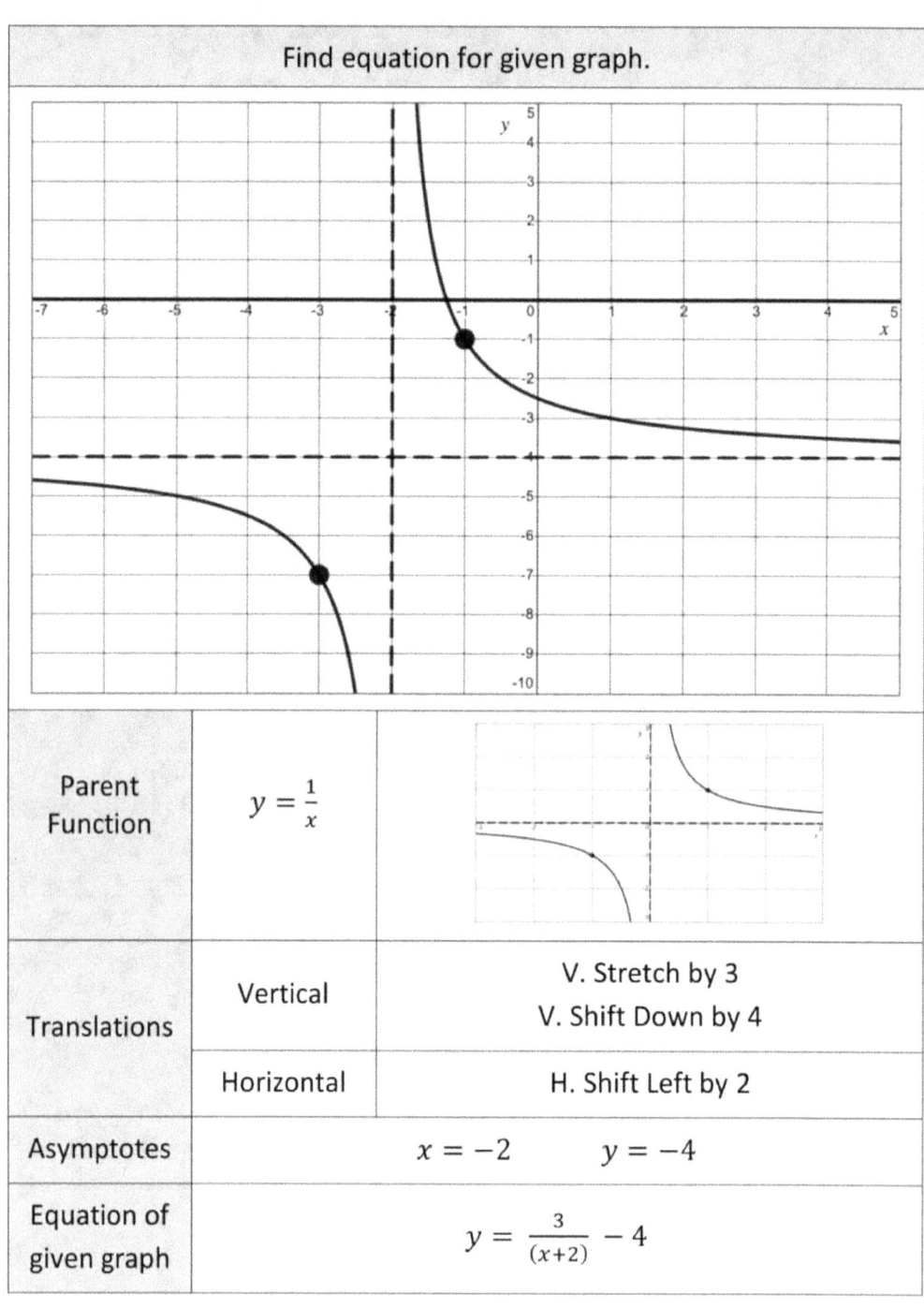

	Find equation for given graph.	
Parent Function	$y = \dfrac{1}{x}$	
Translations	Vertical	V. Stretch by 3 V. Shift Down by 4
	Horizontal	H. Shift Left by 2
Asymptotes	$x = -2$	$y = -4$
Equation of given graph	$y = \dfrac{3}{(x+2)} - 4$	

Rational Functions (part 2)

		Find equation for given graph.
		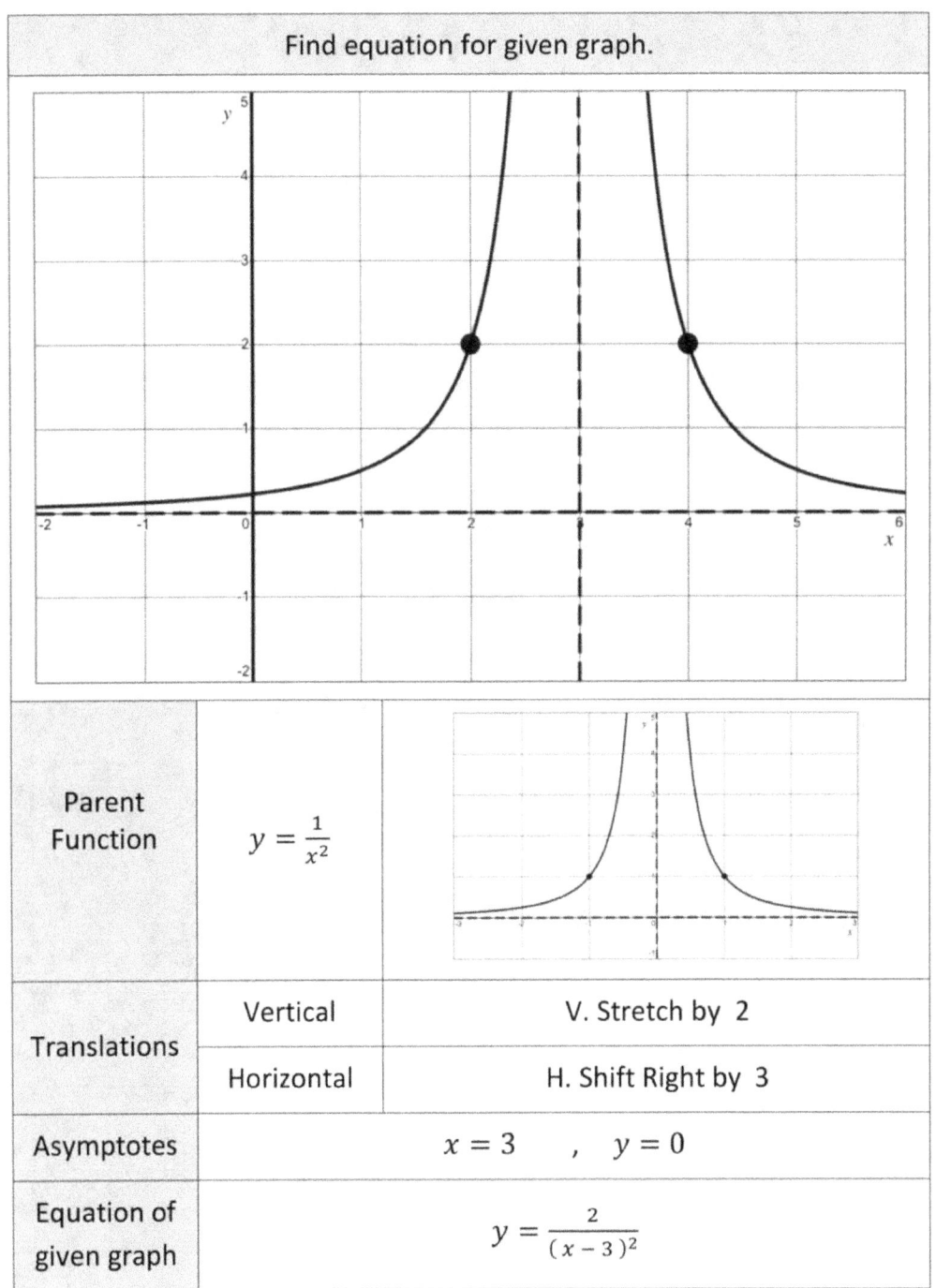
Parent Function	$y = \frac{1}{x^2}$	
Translations	Vertical	V. Stretch by 2
	Horizontal	H. Shift Right by 3
Asymptotes		$x = 3$, $y = 0$
Equation of given graph		$y = \frac{2}{(x-3)^2}$

Step Functions (Greatest Integer <= x)

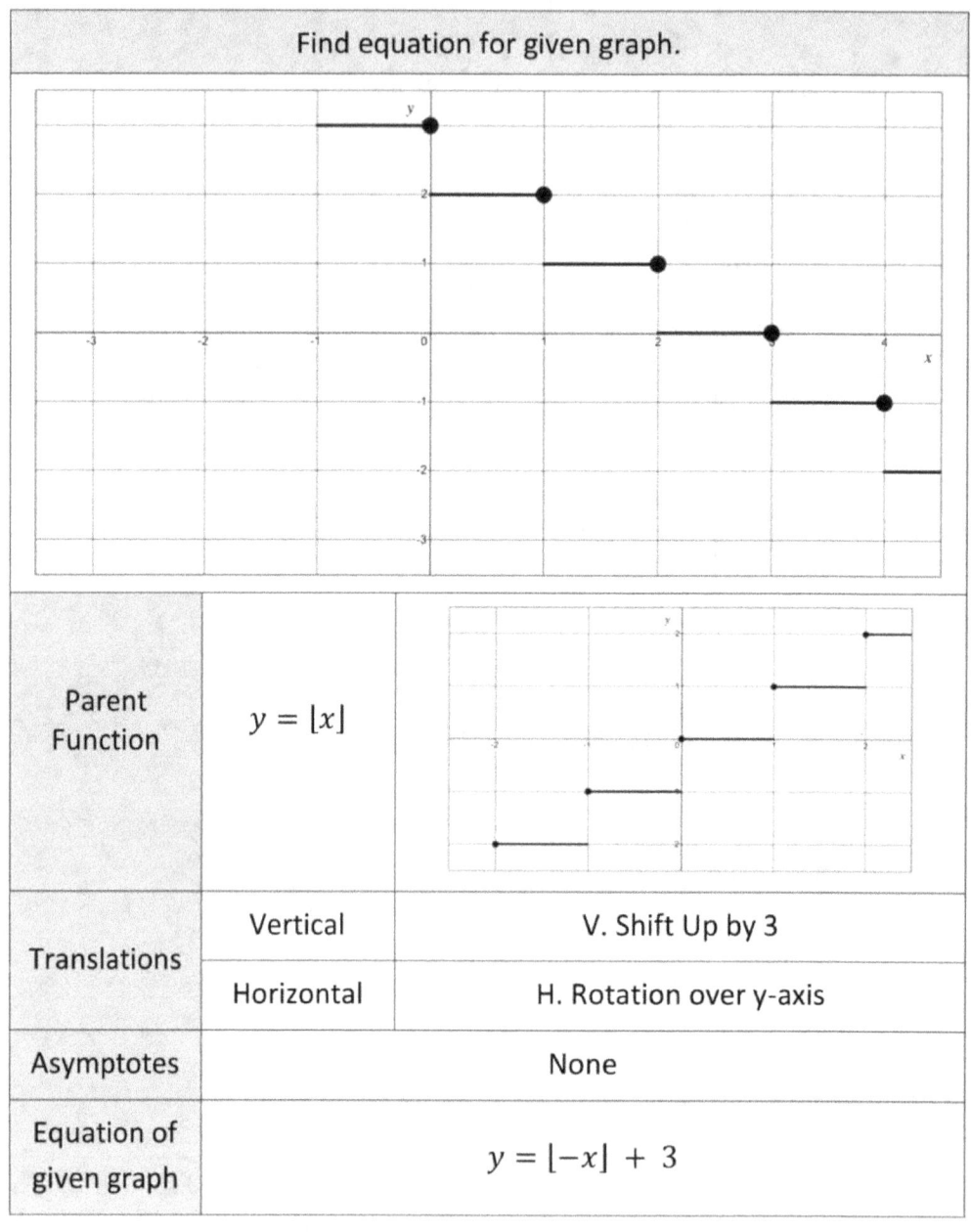

Parent Function	$y = \lfloor x \rfloor$	
Translations	Vertical	V. Shift Up by 3
	Horizontal	H. Rotation over y-axis
Asymptotes		None
Equation of given graph		$y = \lfloor -x \rfloor + 3$

Sine Functions

	Find equation for given graph.
	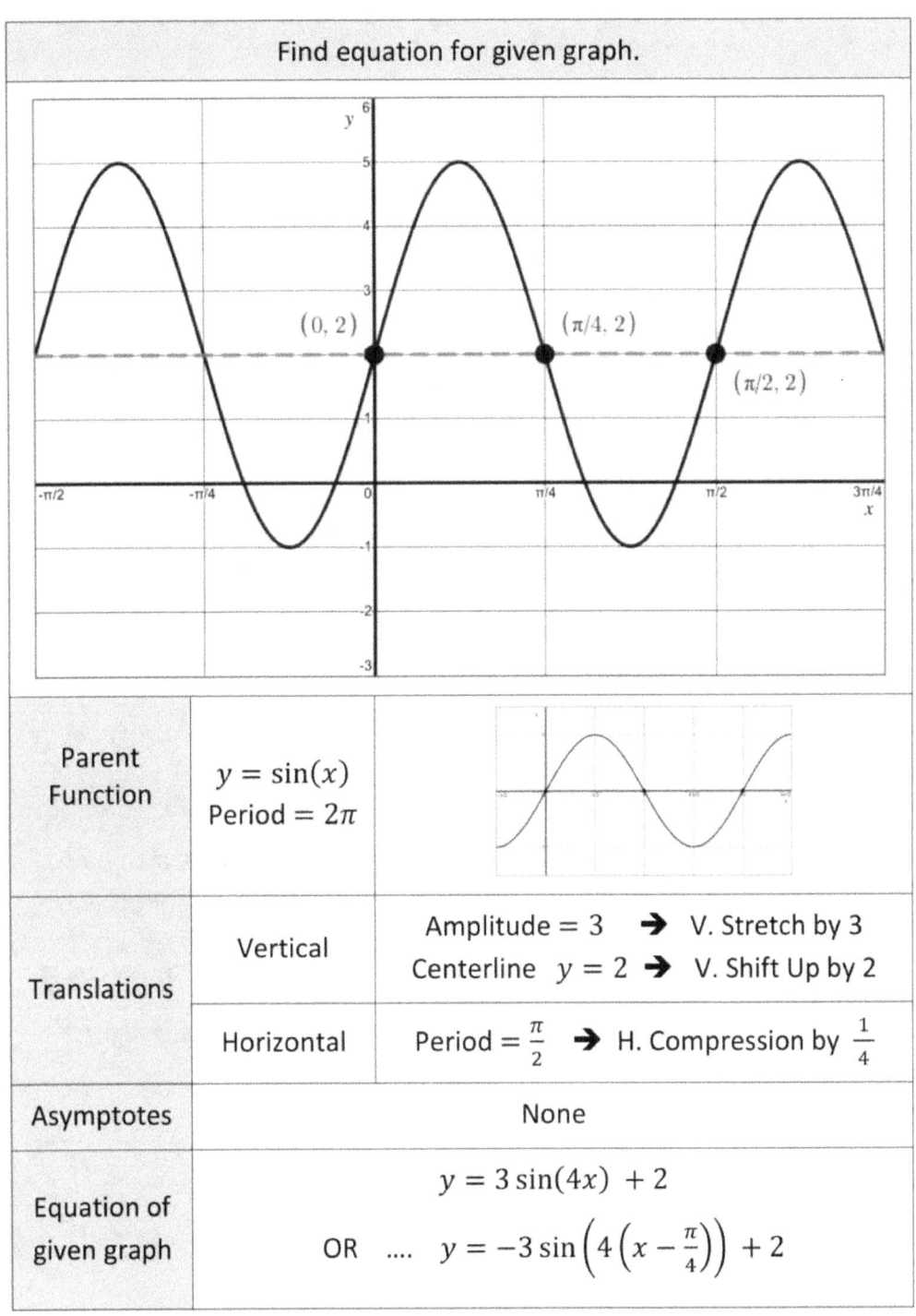

Parent Function	$y = \sin(x)$ Period $= 2\pi$	
Translations	Vertical	Amplitude $= 3$ → V. Stretch by 3 Centerline $y = 2$ → V. Shift Up by 2
	Horizontal	Period $= \frac{\pi}{2}$ → H. Compression by $\frac{1}{4}$
Asymptotes		None
Equation of given graph		$y = 3\sin(4x) + 2$ OR .... $y = -3\sin\left(4\left(x - \frac{\pi}{4}\right)\right) + 2$

Cosine Functions

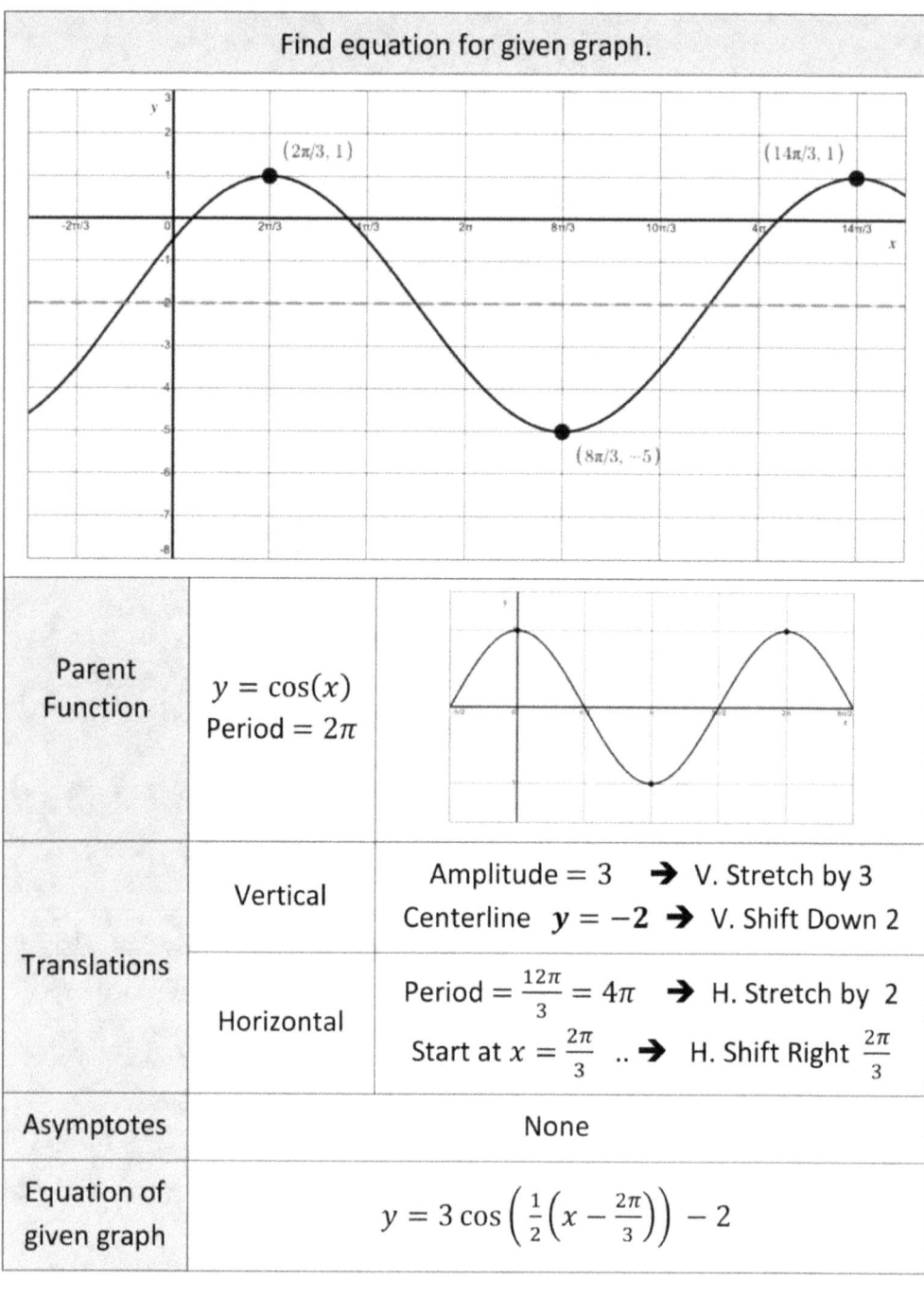

		Find equation for given graph.
Parent Function	$y = \cos(x)$ Period $= 2\pi$	
Translations	Vertical	Amplitude $= 3$ ➔ V. Stretch by 3 Centerline $y = -2$ ➔ V. Shift Down 2
	Horizontal	Period $= \dfrac{12\pi}{3} = 4\pi$ ➔ H. Stretch by 2 Start at $x = \dfrac{2\pi}{3}$ .. ➔ H. Shift Right $\dfrac{2\pi}{3}$
Asymptotes		None
Equation of given graph		$y = 3\cos\left(\dfrac{1}{2}\left(x - \dfrac{2\pi}{3}\right)\right) - 2$

Tangent Functions

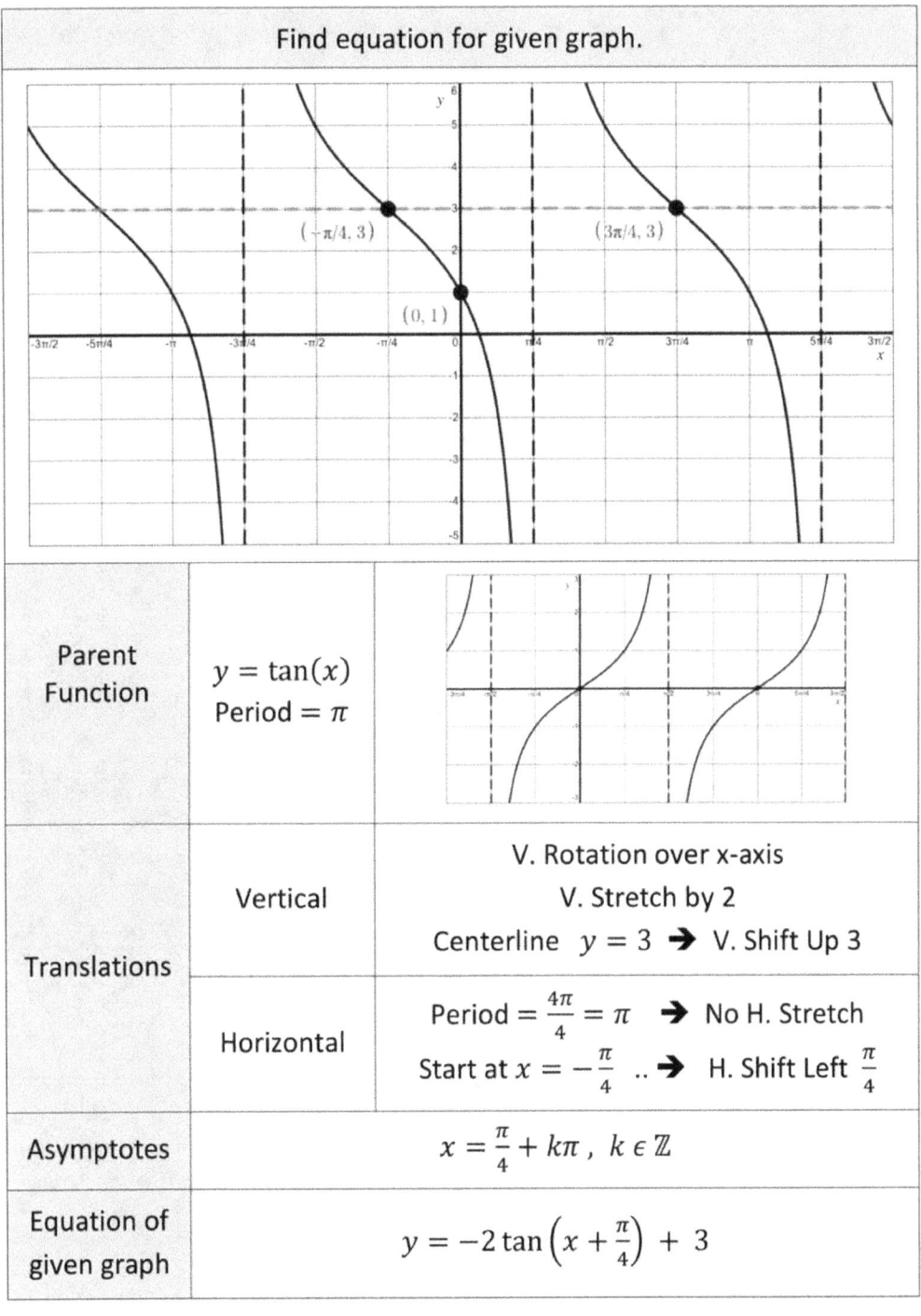

		Find equation for given graph.
Parent Function	$y = \tan(x)$ Period $= \pi$	
Translations	Vertical	V. Rotation over x-axis V. Stretch by 2 Centerline $y = 3$ → V. Shift Up 3
	Horizontal	Period $= \dfrac{4\pi}{4} = \pi$ → No H. Stretch Start at $x = -\dfrac{\pi}{4}$ .. → H. Shift Left $\dfrac{\pi}{4}$
Asymptotes		$x = \dfrac{\pi}{4} + k\pi$, $k \in \mathbb{Z}$
Equation of given graph		$y = -2\tan\left(x + \dfrac{\pi}{4}\right) + 3$

Other Amazon Books By Kathryn Paulk

To see a list of Kathy's books, while on the Amazon website, do a search for "Kathryn Paulk books."

Other Amazon Books by Kathryn Paulk	
One-Page Summaries for Algebra, Geometry, and Pre-Calculus	
Differential Equations With Applications Class Notes With Detailed Examples	
Discrete and Continuous Probability Distributions A Creative Comparison	

Other Amazon Books by Kathryn Paulk		
Big Math for Little Kids Introduction to Numbers	 Workbook	 Solutions
Big Math for Little Kids Introduction to Counting and Fractions by Cooking Breakfast	 Workbook	 Solutions
Big Math for Little Kids Introduction to Fractions by Sharing Things	 Workbook	 Solutions

Other Amazon Books by Kathryn Paulk		
Big Math for Little Kids Learn About Fractions by Baking Cookies	 Workbook	 Solutions
Big Math for Little Kids Adding Big Numbers, Guessing Numbers, And Secret Codes	 Workbook	Solutions
Big Math for Little Kids Learn to Graph by Riding Bikes on Graph Paper	 Workbook	 Solutions

Other Amazon Books by Kathryn Paulk	
BIG Math for Little Kids Help Your Child Learn Math	
Teach Your Child to Swim	B&W Color
Train Your Dog to Use a Litter Box	

www.ingramcontent.com/pod-product-compliance
Lightning Source LLC
Chambersburg PA
CBHW060006230526
45472CB00008B/1971